BEI GRIN MACHT SICH IHR WISSEN BEZAHLT

- Wir veröffentlichen Ihre Hausarbeit, Bachelor- und Masterarbeit

- Ihr eigenes eBook und Buch - weltweit in allen wichtigen Shops

- Verdienen Sie an jedem Verkauf

Jetzt bei www.GRIN.com hochladen und kostenlos publizieren

New Class of Generalized Hermite-Hadamard Type Inequalities

Iram Javed

Bibliografische Information der Deutschen Nationalbibliothek:

Die Deutsche Nationalbibliothek verzeichnet diese Publikation in der Deutschen Nationalbibliografie; detaillierte bibliografische Daten sind im Internet über http://dnb.d-nb.de abrufbar.

ISBN: 9783964877079
Dieses Buch ist auch als E-Book erhältlich.

© GRIN Publishing GmbH
Trappentreustraße 1
80339 München

Druck und Bindung: Books on Demand GmbH, Norderstedt Germany
Gedruckt auf säurefreiem Papier aus verantwortungsvollen Quellen

Das vorliegende Werk wurde sorgfältig erarbeitet. Dennoch übernehmen Autoren und Verlag für die Richtigkeit von Angaben, Hinweisen, Links und Ratschlägen sowie eventuelle Druckfehler keine Haftung.

Das Buch bei GRIN: https://www.grin.com/document/1442888

ABSTRACT

The graphical elegance of fractal theory takes into account the development's achievability and exceptionalism. Due to its fascinating existence in the mathematical fields of sciences, there is a clear association between fractal sets and convexity. In this proposal, we will present generalized convexity and related integral inequalities on a fractal set $\mathbb{R}^{\varpi}$ ($0 < \varpi \leq 1$). In the context of the Beta function, this research presents a new class of generalized Hermite-Hadamard type inequalities. This research contributes significant results of novel versions of fractal Hölder's and Young's inequalities. We derive some general conclusions that capture novel results under investigation. One more remarkable contribution of the study is that two novel auxiliary results along with Trapezoidal and Midpoint type inequalities are provided. Hence, these new results will lead us to generalization of prior results.

TABLE OF CONTENTS

Chapter 1
Preliminaries and Introduction

1 Preliminaries And Introduction

Convex function plays a notable character in the field of both theoretical and applied sciences. The study of convex functions always presents stunning and magnificent sight of the beauty in advanced mathematics. The mathematicians always put potential in this direction as a result, discover and survey a large variety of results that are beneficial and remarkable for applications. This method is effective in dealing with a wide range of problems, the majority of which may be found in both the pure and applied sciences. Convexity also has a finest effect on our daily lives through numerous applications in medicine, industry, business and art. The formulation of inequalities is one of the most important applications of the convex function. Many novel inequalities of various kind of categories related to convex function have been obtained and implemented to other fields of studies can be seen in [1, 2, 3]. In the literature, the (H-H) inequality is highly familiar results. Furthermore, in many fields of science and technology, such as engineering, mathematical statistics, financial economics, and computer science, this inequality has been employed to solve a variety of problems. The definitions and outcomes listed below are considered necessary for our research.

1.1 Convex Functions

Definition 1. [1]

A function $\Phi : I \subset \mathbb{R} \to \mathbb{R}$ is called convex, if

$$\Phi\left(\kappa\zeta_1 + (1-\kappa)\zeta_2\right) \leq \kappa\Phi(\zeta_1) + (1-\kappa)\Phi(\zeta_2), \tag{1}$$

for all $\zeta_1, \zeta_2 \in I$ and $\kappa \in [0,1]$ holds.

1.2 Special functions

Definition 2. [4] Here we recall special function that is Beta function

$$\beta(\varkappa, \varkappa_1) = \frac{\Gamma\varkappa\Gamma\varkappa_1}{\Gamma(\varkappa + \varkappa_1)} = \int_0^1 \kappa^{\varkappa-1}(1-\kappa)^{\varkappa_1-1}\,d\kappa,$$

$\varkappa, \varkappa_1 > 0$.

where $\beta_t(\varkappa, \varkappa_1)$ is incomplete beta function defined by

$$\beta_\kappa(\varkappa, \varkappa_1) = \int_0^\kappa \mathfrak{s}^{(\varkappa-1)}(1-\mathfrak{s})^{(\varkappa_1-1)}\,d\mathfrak{s}, \quad 0 \leq \kappa \leq 1$$

for $\varkappa, \varkappa_1 > 0$.

Where $\Gamma(\varkappa)$ is gamma function defined by

$$\Gamma(\varkappa) := \int_0^\infty E(-\kappa)\kappa^{(\varkappa-1)}\,d\kappa. \tag{2}$$

Fractional calculus widely used in mathematics, engineering and in physics (see [5]). Fractal sets have gotten a lot of interest from scientists and engineers in the last several years. Fractal analysis is a completely new field of study founded on local fractional calculus.It is important to note that fractal analysis presupposes a fundamental improvement in fractal set visualising. Fractals are used in a variety of industries, including music, soil mechanics, photography, and small angle scattering theory. Fractional calculus also used in the field of cryptography. Fractal image compression is one of the most common applications of fractal in software engineering. Yang [6] given the concept of local fractional calculus. Local fractional calculus is a topic that is widely employed in science and technology [7, 8, 9]. Many studies considered the possession of a function on fractal space and developed various fractional calculus concepts by utilising various applications [10, 11]. Authors [12] defined the generalized convex function on the fractal space $\mathbb{R}^{\varpi}$ ($0 < \varpi \leq 1$) and obtained the "generalized (H-H) inequality" for a generalized convex function in the concept of local fractional calculus. Now we use the set $\mathbb{R}^{\varpi}$ to classify the definitions of the local fractional derivative and local fractional integral and so on, recalling the Gao-Yang-Kang notion.

The theory of Yang's fractional sets [6] can be stated as.

For $0 < \varpi \leq 1$, the ϖ-type set of element set are given below:

$\mathbb{Z}^{\varpi}$: The ϖ-type set of integer numbers is used to define the set $\{0^{\varpi}, \pm 1^{\varpi}, \pm 2^{\varpi}, ..., \pm n^{\varpi}, ...\}$.

$\mathbb{Q}^{\varpi}$: The ϖ-type set of rational numbers is used to define the set $\{m^{\varpi} = (u/v)^{\varpi} : u, v \in \mathbb{Z}, v \neq 0\}$.

$\mathbb{J}^{\varpi}$: The ϖ-type set of irrational numbers is used to define the set $\{m^{\varpi} \neq (u/v)^{\varpi} : u, v \in \mathbb{Z}, v \neq 0\}$.

$\mathbb{R}^{\varpi}$: The ϖ-type set of real numbers is used to define the set $\mathbb{R}^{\varpi} = \mathbb{Q}^{\varpi} \cup \mathbb{J}^{\varpi}$.

The following operations hold for u^{ϖ}, v^{ϖ} and τ^{ϖ} belong to the set $\mathbb{R}^{\varpi}$ of real line numbers:

(i) $u^{\varpi} + v^{\varpi}$ and $u^{\varpi} v^{\varpi}$ belong to the set $\mathbb{R}^{\varpi}$;

(ii) $u^{\varpi} + v^{\varpi} = v^{\varpi} + u^{\varpi} = (u+v)^{\varpi} = (u+v)^{\varpi}$;

(iii) $u^{\varpi} + (v^{\varpi} + \tau^{\varpi}) = (u+v)^{\varpi} + \tau^{\varpi}$;

(iv) $u^{\varpi} v^{\varpi} = v^{\varpi} u^{\varpi} = (uv)^{\varpi} = (vu)^{\varpi}$;

(v) $u^{\varpi} (v^{\varpi} \tau^{\varpi}) = (u^{\varpi} v^{\varpi}) \tau^{\varpi}$;

(vi) $u^{\varpi} (v^{\varpi} + \tau^{\varpi}) = u^{\varpi} v^{\varpi} + u^{\varpi} \tau^{\varpi}$;

(vii) $u^{\varpi} + 0^{\varpi} = 0^{\varpi} + u^{\varpi} = u^{\varpi}$ and $u^{\varpi} 1^{\varpi} = 1^{\varpi} u^{\varpi} = u^{\varpi}$.

1.3 Local Fractional Calculus on $\mathbb{R}^{\varpi}$

Definition 3. [6] " A function which is non-differentiable $\Phi : \mathbb{R} \longrightarrow \mathbb{R}^{\varpi}$, $\varkappa_1 \longrightarrow \Phi(\varkappa_1)$ is called to be local fractional continuous at $\varkappa_0$, if for any $\varepsilon \in \mathbb{R}^{+}$, there exist $\delta \in \mathbb{R}^{+}$, such that

$$|\Phi(\varkappa_1) - \Phi(\varkappa_0)| < \varepsilon^{\varpi}$$

holds for $|\varkappa_1 - \varkappa_0| < \delta$, where $\Phi, \delta \in \mathbb{R}$. If $\Phi(\varkappa_1)$ is local fractional which is continuous on the interval $(\vartheta_1, \vartheta_2)$, we denote $\Phi(\theta) \in C_{\varpi}(\vartheta_1, \vartheta_2)$".

Definition 4. [6] "The local fractional derivative of $\Phi(\varkappa)$ of order ϖ at $\varkappa = \varkappa_0$ is defined by

$$\Phi^{(\varpi)}(\varkappa_0) = \varkappa_0 D_\varkappa^\varpi \phi(\varkappa) = \left.\frac{d^{(\varpi)}\Phi(\varkappa)}{d(\varkappa^\varpi)}\right|_{\varkappa=\varkappa_0} = lim_{\varkappa\to\varkappa_0}\frac{\Delta^\varpi(\Phi(\varkappa)-\Phi(\varkappa_0))}{(\varkappa-\varkappa_0)^\varpi}, \quad (3)$$

where $\Delta^\varpi(\Phi(\varkappa) - \Phi(\varkappa_0)) \cong \Gamma(1+\varpi)(\Phi(\varkappa) - \Phi(\varkappa_0))$ and Γ is the well-known gamma function. Let $\Phi^{(\varpi)}(\varkappa) = D_\varkappa^\varpi\Phi(\varkappa)$. If there holds $\Phi^{(k+1)\varpi}(\varkappa) = \overbrace{D_\varkappa^\varpi...D_\varkappa^\varpi}\Phi(\varkappa)$ for any $\varkappa \in I \subseteq \mathbb{R}$, then we denote $\Phi \in D_{(k+1)\varpi}(I)$, where $K \in \mathbb{N}$".

Definition 5. [6] "Let $\Phi \in C_\varpi[\vartheta_1, \vartheta_2]$. In addition , let $P = \kappa_0...\kappa_N$, ($N \in \mathbb{N}$ be one of the interval's partitions $[\vartheta_1, \vartheta_2]$ that satisfies $\vartheta_1 = \kappa_0 < \kappa_1 < ..... < \kappa_{N-1} < \kappa_N = \vartheta_2$. Furthermore, for this partition P, Let $\Delta\kappa := max_{0\leq j\leq N-1}\Delta\kappa_j$, where $\Delta\kappa_j := \kappa_{j+1} - \kappa_j$ and $j = 0,.....,N-1$. Then, Φ is a local fractional integral on the interval $[\vartheta_1, \vartheta_2]$ of order ϖ (denoted by $_{\vartheta_1}I_{\vartheta_2}^{(\varpi)}\Phi$) is defined by

$$_{\vartheta_1}I_{\vartheta_2}^{(\varpi)}\Phi(\kappa) = \frac{1}{\Gamma(1+\varpi)}\int_{\vartheta_1}^{\vartheta_2}\Phi(\kappa)(d\kappa)^\varpi := \frac{1}{\Gamma(1+\varpi)}lim_{\Delta t\to 0}\sum_{j=0}^{N-1}\Phi(\kappa_j)(\Delta\kappa_j)^\varpi, \quad (4)$$

Here, it follows that $_{\vartheta_1}I_{\vartheta_1}^{(\varpi)}\Phi = 0$ if $\vartheta_1 = \vartheta_2$ and $_{\vartheta_1}I_{\vartheta_2}^{(\varpi)}\Phi = -_{\vartheta_2}I_{\vartheta_1}^{(\varpi)}\Phi$ if $\vartheta_1 < \vartheta_2$. If $_{\vartheta_1}I_\varkappa^{(\varpi)}g$ exists for any $\varkappa \in [\vartheta_1, \vartheta_2]$ and a function $g : [\vartheta_1, \vartheta_2] \to \mathbb{R}^\varpi$, after that, we denote $g \in I_\varkappa^{(\varpi)}[\vartheta_1, \vartheta_2]$".
We describe some of the properties of the local fractional calculus that will be necessary for our main conclusions.

Lemma 1.1. *[6] The following identities are correct:*
*(1)("**The local fractional derivative of $\varkappa^{k\varpi}$**"):*

$$\frac{d^\varpi \varkappa^{k\varpi}}{d\varkappa^\varpi} = \frac{\Gamma(1+k\varpi)}{\Gamma(1+(k-1)\varpi)}\varkappa^{(k-1)\varpi} \quad (5)$$

*(2)("**Local fractional integration is anti-differentiation**").*
Assume that $\Phi(\varkappa) = \Psi^{(\varpi)}(\varkappa) \in C_\varpi[\vartheta_1, \vartheta_2]$. Then, we have

$$_{\vartheta_1}I_{\vartheta_2}^{(\varpi)}\Phi(\varkappa) = \Psi(\vartheta_2) - \Psi(\vartheta_1) \quad (6)$$

*(3)("**Local fractional integration by parts**").*
Assume that $\Phi(\varkappa), \Psi(\varkappa) \in D_\varpi[\vartheta_1, \vartheta_2]$ and $\Phi^{(\varpi)}(x), \Psi^{(\varpi)}(\varkappa) \in C_\varpi[\vartheta_1, \vartheta_2]$. Then we have

$$_{\vartheta_1}I_{\vartheta_2}^{(\varpi)}\Phi(\varkappa)\Psi^{(\varpi)}(\varkappa) = \Phi(\varkappa)\Psi(\varkappa)|_{\vartheta_1}^{\vartheta_2} - _{\vartheta_1}I_{\vartheta_2}^{(\varpi)}\Phi^{(\varpi)}(\varkappa)\Psi(\varkappa) \quad (7)$$

*(4)("**The Local fractional definite integrals of $\varkappa^{k\varpi}$**"):*

$$\frac{1}{\Gamma(1+\varpi)}\int_{\vartheta_1}^{\vartheta_2}\varkappa^{k\varpi}(d\varkappa)^\varpi = \frac{\Gamma(1+k\varpi)}{\Gamma(1+(k+1)\varpi)}(\vartheta_2^{(k+1)\varpi} - \vartheta_1^{(k+1)\varpi}), \quad k \in \mathbb{R} \quad (8)$$

For furthur information one can see[13].

1.4 Generalized Convex Function

Definition 6. [12] For any $\zeta_1, \zeta_2 \in I$ and $\kappa \in [0,1]$, if the following inequality

$$\Phi(\kappa\zeta_1 + (1-\kappa)\zeta_2) \leq \kappa^\varpi \Phi(\zeta_1) + (1-\kappa)^\varpi \Phi(\zeta_2), \tag{9}$$

holds, then Φ is called generalized convex function on I, where $\Phi : I \subseteq \mathbb{R} \to \mathbb{R}^\varpi$.

1.5 Generalized Special Functions

Definition 7. [14] For $0 < \varpi \leq 1$ and $\varkappa, \varkappa_1 \in \mathbb{R}^+$, the local Beta function with two parameters $\varkappa$ and $\varkappa_1$ is defined as

$$\beta_\varpi(\varkappa, \varkappa_1) = \frac{\Gamma_\varpi(\varkappa)\Gamma_\varpi(\varkappa_1)}{\Gamma_\varpi(\varkappa + \varkappa_1)} = \int_0^1 \kappa^{(\varkappa-1)\varpi}(1-\kappa)^{(\varkappa_1-1)\varpi}(d\kappa)^\varpi, \tag{10}$$

$\varkappa, \varkappa_1 > 0$. For $\varpi = 1$, the equation 10 gives integral representation of classical Euler beta function $\beta(\varkappa, \varkappa_1)$. Therefore, in this case, $\beta_\varpi(\varkappa, \varkappa_1) = \beta(\varkappa, \varkappa_1)$.

Where for $0 < \varpi \leq 1$ and $\varkappa \in \mathbb{R}^+$, the local gamma function is defined by

$$\Gamma_\varpi(\varkappa) := \frac{1}{\varpi!}\int_0^\infty E_\varpi(-\kappa^\varpi)\kappa^{(\varkappa-1)\varpi}(d\kappa)^\varpi \tag{11}$$

for $\varpi = 1$, the equation 11 gives integral representation of classical Euler gamma function $\Gamma(\varkappa)$. Therefore, in this case, $\Gamma_\varpi(\varkappa) = \Gamma(\varkappa)$.

Definition 8. [14] Here $\beta_{\kappa\varpi}(\varkappa, \varkappa_1)$ is incomplete beta function defined by

$$\beta_{\kappa\varpi}(\varkappa, \varkappa_1) = \int_0^\kappa \mathfrak{s}^{(\varkappa-1)\varpi}(1-\mathfrak{s})^{(\varkappa_1-1)\varpi}(d\mathfrak{s})^\varpi, \qquad 0 \leq \kappa \leq 1$$

for $\varkappa, \varkappa_1 > 0$.

Chapter 2
Back Ground Of The Problem

2 Back Ground Of The Problem

Hermite-Hadamard's inequality is the utmost important and extensively used result involving convex functions. In literature this inequality called classical equation of Hermite-Hadamard inequality. Some recent results about Hermite-Hadamard inequality, reader can see in [15, 16, 17] The Hermite–Hadamard inequality state that,

2.1 Hermite–Hadamard Inequality

Hermite–Hadamard inequality is the best-known result in the literature:

Theorem 2.1. *[2] The Hermite–Hadamard inequality is defined as:*

$$\Phi\left(\frac{\vartheta_1+\vartheta_2}{2}\right) \leq \frac{1}{\vartheta_2-\vartheta_1}\int_{\vartheta_1}^{\vartheta_2}\Phi(\varkappa)d\varkappa \leq \frac{\Phi(\vartheta_1)+\Phi(\vartheta_2)}{2}. \tag{12}$$

Where Φ is a convex function given as $\Phi: I \subseteq \mathbb{R} \to \mathbb{R}$ and $\vartheta_1, \vartheta_2 \in I$ with $\vartheta_1 < \vartheta_2$ In recent years, Hermite–Hadamard inequality has received remarkable attention and there is a memorable collection of refinements and generalizations for convex functions (see [2, 18]) and references therein.

2.2 Jensen-Mercer Inequality

Let $0 < \zeta_1 \leq \zeta_2 \leq ... \leq \zeta_n$ and let $\rho = (\rho_1, \rho_2, ..., \rho_k)$ be non-negative weights such that $\sum_{\lambda=1}^{k} \rho_\lambda = 1$. The famous Jensen inequality [1] in the literature states that if Φ is convex functions on the interval $[\vartheta_1, \vartheta_2]$, then

$$\Phi\left(\sum_{\lambda=1}^{k} \rho_\lambda \zeta_\lambda\right) \leq \left(\sum_{\lambda=1}^{k} \rho_\lambda \Phi(\zeta_\lambda)\right), \tag{13}$$

for all $\zeta_\lambda \in [\vartheta_1, \vartheta_2], \rho_\lambda \in [0,1]$ *and* $(\lambda = 1, 2, ..., k)$.

In 2003, a new variant of Jensen's inequality was introduced by Mercer [19] as: If Φ is a convex function on $[a, b]$, then

$$\Phi\left(\vartheta_1 + \vartheta_2 - \sum_{\lambda=1}^{k} \rho_\lambda \zeta_\lambda\right) \leq \Phi(\vartheta_1) + \Phi(\vartheta_2) - \sum_{\lambda=1}^{k} \rho_\lambda \Phi(\zeta_\lambda), \tag{14}$$

holds for all $\zeta_\lambda \in [\vartheta_1, \vartheta_2], \rho_\lambda \in [0,1]$ *and* $(\lambda = 1, 2, ..., k)$. For more information, one can see [20, 21, 22].

2.3 Generalized Hermite–Hadamard's Inequality

Theorem 2.2. *[12] The Generalized Hermite–Hadamard's Inequality is given as:*

$$\Phi\left(\frac{\vartheta_1+\vartheta_2}{2}\right) \leq \frac{\Gamma(1+\varpi)}{(\vartheta_2-\vartheta_1)^\varpi} {}_{\vartheta_1}I_{\vartheta_2}^{(\varpi)}\Phi(\varkappa) \leq \frac{\Phi(\vartheta_1)+\Phi(\vartheta_2)}{2^\varpi}. \tag{15}$$

Where generalized convex function is $\Phi: I \subseteq \mathbb{R} \to \mathbb{R}^\varpi$ with $\vartheta_1, \vartheta_2 \in I$ and $\vartheta_1 < \vartheta_2$.

2.4 Generalized Jensen-Mercer Inequality

Theorem 2.3. *[12] (Generalized Jensen's Inequality (G.J.I)) Suppose that* $\Phi : \mathbb{R} \to \mathbb{R}^{\varpi}$ *be a generalized convex function on* $I = [\vartheta_1, \vartheta_2]$. *Then for any* $\zeta_{\gamma} \in I$ *and* $\omega_{\gamma} \in [0,1]$ *where* $\gamma = 1, 2, ..., k$ *with* $\sum_{\gamma=1}^{k} = \omega_{\gamma}$, *we have*

$$\Phi\left(\sum_{\gamma=1}^{k} \omega_{\gamma} \zeta_{\gamma} \right) \leq \sum_{\gamma=1}^{k} \omega_{\gamma}^{\sigma} \Phi(\zeta_{\gamma}). \tag{16}$$

Recently, Butt et al. In [23] established the fractal variant of Jensen-Mercer inequality (14) given as:

$$\Phi\left(\vartheta_1 + \vartheta_2 - \sum_{\gamma=1}^{k} \omega_{\gamma} \zeta_{\gamma} \right) \leq \Phi(\vartheta_1) + \Phi(\vartheta_2) - \sum_{\gamma=1}^{k} \omega_{\gamma}^{\sigma} \Phi(\zeta_{\gamma}). \tag{17}$$

Where $I = [\vartheta_1, \vartheta_2]$ be an interval.
Also in the similar paper he gave new refinements and extensions. Some Hermite Jensen-Mercer inequalities are given in [24].

2.5 Generalized Hölder's Inequality

Theorem 2.4. *[6] Let* $\Phi, \Phi \in C_{\varpi}[\vartheta_1, \vartheta_2]$, $\mathfrak{p}$, $\mathfrak{q} > 1$ *are conjugate exponents, then*

$$\frac{1}{\Gamma(1+\varpi)} \int_{\vartheta_1}^{\vartheta_2} |\Phi(\varkappa)\Phi(\varkappa)|(d\varkappa)^{\varpi}$$
$$< \left(\frac{1}{\Gamma(1+\varpi)} \int_{\vartheta_1}^{\vartheta_2} |\Phi(\varkappa)|^{\mathfrak{p}}(d\varkappa)^{\varpi} \right)^{\frac{1}{\mathfrak{p}}} \left(\frac{1}{\Gamma(1+\varpi)} \int_{\vartheta_1}^{\vartheta_2} |\Phi(\varkappa)|^{\mathfrak{q}}(d\varkappa)^{\varpi} \right)^{\frac{1}{\mathfrak{q}}}. \tag{18}$$

2.6 Generalized Young's Inequality

Theorem 2.5. *[25] Let* $\Phi, \Phi \in C_{\varpi}[\vartheta_1, \vartheta_2]$, $\mathfrak{p}$, $\mathfrak{q} > 1$ *are conjugate exponents, then*

$$\frac{1}{\Gamma(1+\varpi)} \int_{\vartheta_1}^{\vartheta_2} |\Phi(\varkappa)\Phi(\varkappa)|(d\varkappa)^{\varpi}$$
$$\leq \frac{1}{\mathfrak{p}} \left(\frac{1}{\Gamma(1+\varpi)} \int_{\vartheta_1}^{\vartheta_2} |\Phi(\varkappa)|^{\mathfrak{p}}(d\varkappa)^{\varpi} \right) + \frac{1}{\mathfrak{q}} \left(\frac{1}{\Gamma(1+\varpi)} \int_{\vartheta_1}^{\vartheta_2} |\Phi(\varkappa)|^{\mathfrak{q}}(d\varkappa)^{\varpi} \right). \tag{19}$$

Chapter 3
Main Results

3 Main Results

3.1 Result's for Fractal Trapezoid Inequalities

We start this section by proving following Lemma for Φ^{ϖ} which are further utilized to formulate new inequalities.

Lemma 3.1. *Let $I \subset \mathbb{R}$ be an interval, $\Phi : I^o \subset \mathbb{R} \to \mathbb{R}^{\varpi}$ (I^o is the interior of I) such that $\Phi \in D_{\varpi}(I^o)$ and $\Phi^{\varpi} \in C_{\varpi}[\vartheta_1, \vartheta_2]$ for $\vartheta_1, \vartheta_2 \in I^o$ with $\vartheta_1 < \vartheta_2$, we have the identity*

$$
\frac{\Phi(\vartheta_1) + \Phi(\vartheta_2)}{2^{\varpi}} \beta_{\varpi}(m_1, m_2) - \frac{\Gamma(\varpi + 1)}{2^{\varpi}} \frac{1}{(\vartheta_2 - \vartheta_1)^{(m_1 + m_2 - 1)\varpi}} {}_{\vartheta_1} I_{\vartheta_2}^{(\varpi)} \Omega \Phi
$$
$$
= \frac{(\vartheta_2 - \vartheta_1)^{\varpi}}{2^{\varpi}} \frac{1}{\Gamma(1 + \varpi)} \int_0^1 \beta_{\kappa\varpi}(m_1, m_2)
$$
$$
\times [\Phi^{(\varpi)}(\kappa\vartheta_2 + (1 - \kappa)\vartheta_1) - \Phi^{(\varpi)}(\kappa\vartheta_1 + (1 - \kappa)\vartheta_2)](d\kappa)^{\varpi} \tag{20}
$$

where $\beta_{\kappa\varpi}(m_1, m_2)$ is incomplete beta function defined by

$$
\beta_{\kappa\varpi}(m_1, m_2) = \int_0^{\kappa} \mathfrak{s}^{(m_1 - 1)\varpi}(1 - \mathfrak{s})^{(m_2 - 1)\varpi}(d\mathfrak{s})^{\varpi}, \qquad 0 \leq \kappa \leq 1
$$

and

$$
\Omega(\varkappa) = (\varkappa - \vartheta_1)^{(m_1 - 1)\varpi}(\vartheta_2 - \varkappa)^{(m_2 - 1)\varpi} + (\varkappa - \vartheta_1)^{(m_2 - 1)\varpi}(\vartheta_2 - \varkappa)^{(m_1 - 1)\varpi} \tag{21}
$$

for $m_1, m_2 > 0$.

Proof. Here, we consider following integrals and using integration by parts, then we have

$$
\Phi_1 = \frac{1}{\Gamma(1 + \varpi)} \int_0^1 \beta_{t\varpi}(m_1, m_2) \Phi^{(\varpi)}(\kappa\vartheta_2 + (1 - \kappa)\vartheta_1)(d\kappa)^{\varpi}
$$
$$
= \frac{\beta_{\varpi}(m_1, m_2)\Phi(\vartheta_2)}{(\vartheta_2 - \vartheta_1)^{\varpi}} - \frac{\Gamma(1 + \varpi)}{(\vartheta_2 - \vartheta_1)^{\varpi}}
$$
$$
\times \left(\frac{1}{\Gamma(1 + \varpi)} \int_0^1 \Phi(\kappa\vartheta_2 + (1 - \kappa)\vartheta_1) \kappa^{(m_1 - 1)\varpi}(1 - \kappa)^{(m_2 - 1)\varpi}(d\kappa)^{\varpi} \right)
$$
$$
= \frac{\beta_{\varpi}(m_1, m_2)\Phi(\vartheta_2)}{(\vartheta_2 - \vartheta_1)^{\varpi}} - \frac{\Gamma(1 + \varpi)}{(\vartheta_2 - \vartheta_1)^{\varpi}} \frac{1}{(\vartheta_2 - \vartheta_1)^{(m_1 + m_2 - 1)\varpi}}
$$
$$
\times \left(\frac{1}{\Gamma(1 + \varpi)} \int_a^b (\varkappa - \vartheta_1)^{(m_1 - 1)\varpi}(\vartheta_2 - \varkappa)^{(m_2 - 1)\varpi} \Phi(\varkappa)(d\varkappa)^{\varpi} \right).
$$

And similarly, we obtain

$$\Phi_2 = \frac{1}{\Gamma(1+\varpi)} \int_0^1 \beta_{t\varpi}(m_1, m_2) \Phi^{(\varpi)}(\kappa\vartheta_1 + (1-\kappa)\vartheta_2)(d\kappa)^\varpi$$

$$= -\frac{\beta_\varpi(m_1, m_2)\Phi(\vartheta_1)}{(\vartheta_2 - \vartheta_1)^\varpi} + \frac{\Gamma(1+\varpi)}{(\vartheta_2 - \vartheta_1)^\varpi}$$

$$\times \left(\frac{1}{\Gamma(1+\varpi)} \int_0^1 \Phi(\kappa\vartheta_1 + (1-\kappa)\vartheta_2) \kappa^{(m_1-1)\varpi}(1-\kappa)^{(m_2-1)\varpi}(d\kappa)^\varpi \right)$$

$$= -\frac{\beta_\varpi(m_1, m_2)\Phi(\vartheta_1)}{(\vartheta_2 - \vartheta_1)^\varpi} + \frac{\Gamma(1+\varpi)}{(\vartheta_2 - \vartheta_1)^\varpi}$$

$$\times \frac{1}{(\vartheta_2 - \vartheta_1)^{(m_1+m_2-1)\varpi}} \left(\frac{1}{\Gamma(1+\varpi)} \int_a^b (\varkappa - \vartheta_1)^{(m_2-1)\varpi}(\vartheta_2 - \varkappa)^{(m_1-1)\varpi}\Phi(\varkappa)(d\varkappa)^\varpi \right).$$

If we subtract Φ_2 from Φ_1 and multiply by $\frac{(\vartheta_2 - \vartheta_1)^\varpi}{2^\varpi}$ then the proof is completed. $\qquad \square$

Remark 3.2. *Substituting $\varpi = 1$ in Lemma 43 then:*

$$\frac{\Phi(\vartheta_1) + \Phi(\vartheta_2)}{2} \beta(m_1, m_2) - \frac{1}{2} \frac{1}{(\vartheta_2 - \vartheta_1)^{(m_1+m_2-1)}} \int_{\vartheta_1}^{\vartheta_2} \Omega(\varkappa)\Phi(\varkappa)d\varkappa$$

$$= \frac{(\vartheta_2 - \vartheta_1)}{2} \int_0^1 \beta_\kappa(m_1, m_2)[\Phi'(\kappa\vartheta_2 + (1-\kappa)\vartheta_1) - \Phi'(\kappa\vartheta_1 + (1-\kappa)\vartheta_2)]d\kappa \qquad (22)$$

holds. which is proved by M. Z. Sarikaya in [26].
Here we have some scenarios as:

- *If in equation (46), we get $m_1 = 1$, $m_2 = \varpi$, (or $m_2 = 1$, $m_1 = \varpi$) then we have Lemma (2) proved by Sarikaya et al. in [27].*

- *If in equation (46), we get $m_1 = 1$, $m_2 = \frac{\varpi}{k}$, (or $m_2 = 1, m_1 = \frac{\varpi}{k}$) then we have*

$$\frac{\Phi(\vartheta_1) + \Phi(\vartheta_2)}{2} - \frac{\Gamma_k(\varpi + k)}{(\vartheta_2 - \vartheta_1)^{\frac{\varpi}{k}}} \left[J_{\vartheta_1^+, k}^\varpi \Phi(\vartheta_2) + J_{\vartheta_2^-, k}^\varpi \Phi(\vartheta_1) \right]$$

$$= \frac{(\vartheta_2 - \vartheta_1)}{2} \int_0^1 [(1-\kappa)^{\frac{\varpi}{k}} - \kappa^{\frac{\varpi}{k}}]\Phi'(\kappa\vartheta_1 + (1-\kappa)\vartheta_2)d\kappa \qquad (23)$$

which is proved by Hussain in [28].

- *If in Lemma 43, we get $\varpi = m_2 = m_1 = 1$, then we have Lemma (2.1) proved by Dragomir et. al. in [29].*

Remark 3.3. *In Lemma 43, when we change the variable then the equality (43) reduces to*

$$\frac{\Phi(\vartheta_1) + \Phi(\vartheta_2)}{2^\varpi} \beta_\varpi(m_1, m_2) - \frac{\Gamma(\varpi + 1)}{2^\varpi} \frac{1}{(\vartheta_2 - \vartheta_1)^{(m_1+m_2-1)\varpi}} \vartheta_1 I_{\vartheta_2}^{(\varpi)} \Omega\Phi$$

$$= \frac{(\vartheta_2 - \vartheta_1)^\varpi}{2^\varpi} \frac{1}{\Gamma(1+\varpi)} \int_0^1 [\beta_{\kappa\varpi}(m_1, m_2) - \beta_{(1-\kappa)\varpi}(m_1, m_2)]\Phi^\varpi(\kappa\vartheta_2 + (1-\kappa)\vartheta_1)(d\kappa)^\varpi.$$

$$(24)$$

Theorem 3.4. *Let* Φ *be defined as in Lemma 3.14 and if the mapping* $|\Phi^{(\varpi)}|$ *is convex on* $[\vartheta_1, \vartheta_2]$, *then:*

$$\left| \frac{\Phi(\vartheta_1) + \Phi(\vartheta_2)}{2^{\varpi}} \beta_{\varpi}(m_1, m_2) - \frac{\Gamma(\varpi+1)}{2^{\varpi}} \frac{1}{(\vartheta_2 - \vartheta_1)^{(m_1+m_2-1)\varpi}} {}_{\vartheta_1} I_{\vartheta_2}^{\varpi} \Omega \Phi \right|$$

$$\leq \frac{(\vartheta_2 - \vartheta_1)^{\varpi}}{2^{\varpi}} \left[|\Phi^{(\varpi)}(\vartheta_1)| + |\Phi^{(\varpi)}(\vartheta_2)| \right] \tag{25}$$

$$\times \left(\frac{1}{\Gamma(1+\varpi)} \int_0^{\frac{1}{2}} |\beta_{\kappa\varpi}(m_1, m_2) - \beta_{(1-\kappa)\varpi}(m_1, m_2)| (d\kappa)^{\varpi} \right)$$

for $m_1, m_2 > 0$.

Proof. Utilizing Lemma 43 with convexity of $|\Phi^{(\varpi)}|$,

$$\left| \frac{\Phi(\vartheta_1) + \Phi(\vartheta_2)}{2^{\varpi}} \beta_{\varpi}(m_1, m_2) - \frac{\Gamma(\varpi+1)}{2^{\varpi}} \frac{1}{(\vartheta_2 - \vartheta_1)^{(m_1+m_2-1)\varpi}} {}_{\vartheta_1} I_{\vartheta_2}^{\varpi} \Omega \Phi \right|$$

$$\leq \frac{(\vartheta_2 - \vartheta_1)^{\varpi}}{2^{\varpi}} \left(\frac{1}{\Gamma(1+\varpi)} \int_0^1 \left| \beta_{\kappa\varpi}(m_1, m_2) - \beta_{(1-\kappa)\varpi}(m_1, m_2) \right| \right.$$

$$\times \left. \left| \Phi^{(\varpi)}(\kappa\vartheta_2 + (1-\kappa)\vartheta_1 \right| (d\kappa)^{\varpi} \right)$$

$$\leq \frac{(\vartheta_2 - \vartheta_1)^{\varpi}}{2^{\varpi}} \frac{1}{\Gamma(1+\varpi)} \int_0^{\frac{1}{2}} \left[\beta_{\kappa\varpi}(m_1, m_2) - \beta_{(1-\kappa)\varpi}(m_1, m_2) \right]$$

$$\times \left[\kappa^{\varpi} |\Phi^{(\varpi)}(\vartheta_2)| + (1-\kappa)^{\varpi} |\Phi^{(\varpi)}(\vartheta_1)| \right] (d\kappa)^{\varpi}$$

$$+ \frac{(\vartheta_2 - \vartheta_1)^{\varpi}}{2^{\varpi}} \frac{1}{\Gamma(1+\varpi)} \int_{\frac{1}{2}}^1 \left[\beta_{(1-\kappa)\varpi}(m_1, m_2) - \beta_{\kappa\varpi}(m_1, m_2) \right]$$

$$\left[\kappa^{\varpi} |\Phi^{(\varpi)}(\vartheta_2)| + (1-\kappa)^{\varpi} |\Phi^{(\varpi)}(\vartheta_1)| \right] (d\kappa)^{\varpi}$$

$$\leq \frac{(\vartheta_2 - \vartheta_1)^{\varpi}}{2^{\varpi}} \left[|\Phi^{(\varpi)}(\vartheta_1)| + |\Phi^{(\varpi)}(\vartheta_2)| \right]$$

$$\times \left(\frac{1}{\Gamma(1+\varpi)} \int_0^{\frac{1}{2}} \left| \beta_{\kappa\varpi}(m_1, m_2) - \beta_{(1-\kappa)\varpi}(m_1, m_2) \right| (d\kappa)^{\varpi} \right)$$

which completes the proof. $\qquad\square$

Remark 3.5. *If we set* $\varpi = 1$ *in Theorem 3.18 then following result holds:,*

$$\left| \frac{\Phi(\vartheta_1) + \Phi(\vartheta_2)}{2} \beta(m_1, m_2) - \frac{1}{2} \frac{1}{(\vartheta_2 - \vartheta_1)^{(m_1+m_2-1)}} \int_{\vartheta_1}^{\vartheta_2} \Omega(\varkappa)\Phi(\varkappa)d\varkappa \right|$$

$$\leq \frac{(\vartheta_2 - \vartheta_1)}{2} \left[|\Phi'(\vartheta_1)| + |\Phi'(\vartheta_2)| \right] \int_0^{\frac{1}{2}} \left| \beta_{\kappa}(m_1, m_2) - \beta_{(1-\kappa)}(m_1, m_2) \right| d\kappa, \tag{26}$$

which is proved by M. Z. Sarikaya in [26].
Here we have other scenarios as:

- *If in equation (53), we get* $m_1 = 1, m_2 = \varpi$, *(or* $m_2 = 1$, $m_1 = \varpi$*), then we have*

$$\left| \frac{\Phi(\vartheta_1) + \Phi(\vartheta_2)}{2} - \frac{\Gamma(\varpi+1)}{(\vartheta_2 - \vartheta_1)^\varpi} \left[J^\varpi_{\vartheta_1^+} \Phi(\vartheta_2) + J^\varpi_{\vartheta_2^-} \Phi(\vartheta_1) \right] \right|$$
$$= \frac{(\vartheta_2 - \vartheta_1)}{2(\varpi+1)} \left(1 - \frac{1}{2^\varpi} \right) [\Phi'(\vartheta_2) + \Phi'(\vartheta_1)], \tag{27}$$

which is proved by M. Iqbal in [30].

- *If in equation (53), we get* $m_1 = 1$, $m_2 = \frac{\varpi}{k}$, *(or* $m_2 = 1, m_1 = \frac{\varpi}{k}$*) then we have*

$$\left| \frac{\Phi(\vartheta_1) + \Phi(\vartheta_2)}{2} - \frac{\Gamma_k(\varpi+k)}{2(\vartheta_2 - \vartheta_1)^\varpi} \left[J^\varpi_{\vartheta_1^+, k} \Phi(\vartheta_2) + J^\varpi_{\vartheta_2^-, k} \Phi(\vartheta_1) \right] \right|$$
$$= \frac{(\vartheta_2 - \vartheta_1)}{(\frac{\varpi}{k}+1)} \left(1 - \frac{1}{2^{\frac{\varpi}{k}}} \right) [\Phi'(\vartheta_2) + \Phi'(\vartheta_1)], \tag{28}$$

which is proved by Hussain et. al. in [28].

- *If in Theorem 3.18, we get* $\varpi = m_1 = m_2 = 1$, *then we have Theorem (2.2) proved by Dragomir et. al. in [29].*

Theorem 3.6. *Let* Φ *be defined as in Lemma 3.14 and if the mapping* $|\Phi^{(\varpi)}|^q$, $q > 1$ *is convex on* $[\vartheta_1, \vartheta_2]$, *then we have the following inequality:*

$$\left| \frac{\Phi(\vartheta_1) + \Phi(\vartheta_2)}{2^\varpi} \beta_\varpi(m_1, m_2) - \frac{\Gamma(\varpi+1)}{2^\varpi} \frac{1}{(\vartheta_2 - \vartheta_1)^{(m_1+m_2-1)\varpi}} {}_{\vartheta_1} I^\varpi_{\vartheta_2} \Omega \Phi \right|$$
$$\leq \frac{(\vartheta_2 - \vartheta_1)^\varpi}{2^\varpi} \left[\frac{1}{\mathfrak{p}} \left(\frac{1}{\Gamma(1+\varpi)} \int_0^1 |\beta_{\kappa\varpi}(m_1, m_2) - \beta_{(1-\kappa)\varpi}(m_1, m_2)|^{\mathfrak{p}} (d\kappa)^\varpi \right) \right.$$
$$\left. + \frac{1}{\mathfrak{q}} \left(\frac{\Gamma(1+\varpi)(|\Phi^{(\varpi)}(\vartheta_2)|^{\mathfrak{q}} - |\Phi^{(\varpi)}(\vartheta_1)|^{\mathfrak{q}})}{\Gamma(1+2\varpi)} + \frac{|\Phi^{(\varpi)}(\vartheta_1)|^{\mathfrak{q}}}{\Gamma(\varpi+1)} \right) \right] \tag{29}$$

for $m_1, m_2 > 0$, *where* $\frac{1}{\mathfrak{p}} + \frac{1}{\mathfrak{q}} = 1$.

Proof. Utilizing Lemma 43 and Young's inequality with convexity of $|\Phi^{(\varpi)}|^q$, then we have

$$\left| \frac{\Phi(\vartheta_1) + \Phi(\vartheta_2)}{2^{\varpi}} \beta_{\varpi}(\mathfrak{m}_1, \mathfrak{m}_2) - \frac{\Gamma(\varpi+1)}{2^{\varpi}} \frac{1}{(\vartheta_2 - \vartheta_1)^{(\mathfrak{m}_1+\mathfrak{m}_2-1)\varpi}} {}_{\vartheta_1} I^{\varpi}_{\vartheta_2} \Omega\Phi \right|$$

$$\leq \frac{(\vartheta_2 - \vartheta_1)^{\varpi}}{2^{\varpi}} \left[\frac{1}{\mathfrak{p}} \left(\frac{1}{\Gamma(1+\varpi)} \int_0^1 \left| \beta_{\kappa\varpi}(\mathfrak{m}_1, \mathfrak{m}_2) - \beta_{(1-\kappa)\varpi}(\mathfrak{m}_1, \mathfrak{m}_2) \right|^{\mathfrak{p}} (d\kappa)^{\varpi} \right) \right.$$

$$\left. + \frac{1}{\mathfrak{q}} \left(\frac{1}{\Gamma(1+\varpi)} \int_0^1 \left| \Phi^{(\varpi)}(\kappa\vartheta_2 + (1-\kappa)\vartheta_1) \right|^{\mathfrak{q}} (d\kappa)^{\varpi} \right) \right]$$

$$\leq \frac{(\vartheta_2 - \vartheta_1)^{\varpi}}{2^{\varpi}} \left[\frac{1}{\mathfrak{p}} \left(\frac{1}{\Gamma(1+\varpi)} \int_0^1 |\beta_{\kappa\varpi}(\mathfrak{m}_1, \mathfrak{m}_2) - \beta_{(1-\kappa)\varpi}(\mathfrak{m}_1, \mathfrak{m}_2)|^{\mathfrak{p}} (d\kappa)^{\varpi} \right) \right.$$

$$\left. + \frac{1}{\mathfrak{q}} \left(\frac{1}{\Gamma(1+\varpi)} \int_0^1 \left(\kappa^{\varpi} |\Phi^{(\varpi)}(\vartheta_2)|^{\mathfrak{q}} + (1-\kappa)^{\varpi} |\Phi^{(\varpi)}(\vartheta_1)|^{\mathfrak{q}} \right) (d\kappa)^{\varpi} \right) \right]$$

$$\leq \frac{(\vartheta_2 - \vartheta_1)^{\varpi}}{2^{\varpi}} \left[\frac{1}{\mathfrak{p}} \left(\frac{1}{\Gamma(1+\varpi)} \int_0^1 |\beta_{\kappa\varpi}(\mathfrak{m}_1, \mathfrak{m}_2) - \beta_{(1-\kappa)\varpi}(\mathfrak{m}_1, \mathfrak{m}_2)|^{\mathfrak{p}} (d\kappa)^{\varpi} \right) \right.$$

$$\left. + \frac{1}{\mathfrak{q}} \left(\frac{\Gamma(1+\varpi)(|\Phi^{(\varpi)}(\vartheta_2)|^{\mathfrak{q}} - |\Phi^{(\varpi)}(\vartheta_1)|^{\mathfrak{q}})}{\Gamma(1+2\varpi)} + \frac{|\Phi^{(\varpi)}(\vartheta_1)|^{\mathfrak{q}}}{\Gamma(1+\varpi)} \right) \right]$$

which completes the proof of the Theorem 3.6. $\qquad\square$

Remark 3.7. *If we set $\varpi = 1$ in Theorem 3.6, then the following inequality holds:*

$$\left| \frac{\Phi(\vartheta_1) + \Phi(\vartheta_2)}{2} \beta(\mathfrak{m}_1, \mathfrak{m}_2) - \frac{1}{2} \frac{1}{(\vartheta_2 - \vartheta_1)^{(\mathfrak{m}_1+\mathfrak{m}_2-1)}} \int_{\vartheta_1}^{\vartheta_2} \Omega(\varkappa)\Phi(\varkappa)d\varkappa \right|$$

$$\leq \frac{(\vartheta_2 - \vartheta_1)}{2} \left[\frac{1}{\mathfrak{p}} \left(\int_0^1 |\beta_{\kappa}(\mathfrak{m}_1, \mathfrak{m}_2) - \beta_{(1-\kappa)}(\mathfrak{m}_1, \mathfrak{m}_2)|^{\mathfrak{p}} d\kappa \right) + \frac{1}{\mathfrak{q}} \left(\frac{|\Phi'(\vartheta_2)|^{\mathfrak{q}} + |\Phi'(\vartheta_1)|^{\mathfrak{q}}}{2} \right) \right] \tag{30}$$

which is a new inequality in literature.

Theorem 3.8. *Let Φ be defined as in Lemma 3.14 and if the mapping $|\Phi^{(\varpi)}|^q$, $q > 1$ is convex on $[\vartheta_1, \vartheta_2]$, then we have the following inequality:*

$$\left| \frac{\Phi(\vartheta_1) + \Phi(\vartheta_2)}{2^{\varpi}} \beta_{\varpi}(\mathfrak{m}_1, \mathfrak{m}_2) - \frac{\Gamma(\varpi+1)}{2^{\varpi}} \frac{1}{(\vartheta_2 - \vartheta_1)^{(\mathfrak{m}_1+\mathfrak{m}_2-1)\varpi}} {}_{\vartheta_1} I^{\varpi}_{\vartheta_2} \Omega\Phi \right|$$

$$\leq \frac{(\vartheta_2 - \vartheta_1)^{\varpi}}{2^{\varpi}} \left(\frac{1}{\Gamma(1+\varpi)} \int_0^1 |\beta_{\kappa\varpi}(\mathfrak{m}_1, \mathfrak{m}_2) - \beta_{(1-\kappa)\varpi}(\mathfrak{m}_1, \mathfrak{m}_2)|^{\mathfrak{p}} (d\kappa)^{\varpi} \right)^{\frac{1}{\mathfrak{p}}}$$

$$\times \left(\frac{\Gamma(1+\varpi)(|\Phi^{(\varpi)}(\vartheta_2)|^{\mathfrak{q}} - |\Phi^{(\varpi)}(\vartheta_1)|^{\mathfrak{q}})}{\Gamma(1+2\varpi)} + \frac{|\Phi^{(\varpi)}(\vartheta_1)|^{\mathfrak{q}}}{\Gamma(\varpi+1)} \right)^{\frac{1}{\mathfrak{q}}} \tag{31}$$

for $\mathfrak{m}_1, \mathfrak{m}_2 > 0$, where $\frac{1}{\mathfrak{p}} + \frac{1}{\mathfrak{q}} = 1$.

Proof. Using Hölder's inequality in Lemma 43 with convexity of $|\Phi^{(\varpi)}|^q$ then we have

$$\left| \frac{\Phi(\vartheta_1) + \Phi(\vartheta_2)}{2\varpi} \mathcal{B}_\varpi(m_1, m_2) - \frac{\Gamma(\varpi+1)}{2\varpi} \frac{1}{(\vartheta_2 - \vartheta_1)^{(m_1+m_2-1)\varpi}} {}_{\vartheta_1} I^\varpi_{\vartheta_2} \Omega\Phi \right|$$

$$\leq \frac{(\vartheta_2 - \vartheta_1)^\varpi}{2\varpi} \left(\frac{1}{\Gamma(1+\varpi)} \int_0^1 |\mathcal{B}_{\kappa\varpi}(m_1, m_2) - \mathcal{B}_{(1-\kappa)\varpi}(m_1, m_2)|^p (d\kappa)^\varpi \right)^{\frac{1}{p}}$$

$$\times \left(\frac{1}{\Gamma(1+\varpi)} \int_0^1 \left| \Phi^{(\varpi)} \left(\kappa\vartheta_2 + (1-\kappa)\vartheta_1 \right) \right|^q (d\kappa)^\varpi \right)^{\frac{1}{q}}$$

$$\leq \frac{(\vartheta_2 - \vartheta_1)^\varpi}{2\varpi} \left(\frac{1}{\Gamma(1+\varpi)} \int_0^1 |\mathcal{B}_{\kappa\varpi}(m_1, m_2) - \mathcal{B}_{(1-\kappa)\varpi}(m_1, m_2)|^p (d\kappa)^\varpi \right)^{\frac{1}{p}}$$

$$\times \left(\frac{1}{\Gamma(1+\varpi)} \int_0^1 \left(\kappa^\varpi |\Phi^{(\varpi)}(\vartheta_2)|^q + (1-\kappa)^\varpi |\Phi^{(\varpi)}(\vartheta_1)|^q \right) (d\kappa)^\varpi \right)^{\frac{1}{q}}$$

$$\leq \frac{(\vartheta_2 - \vartheta_1)^\varpi}{2\varpi} \left(\frac{1}{\Gamma(1+\varpi)} \int_0^1 |\mathcal{B}_{\kappa\varpi}(m_1, m_2) - \mathcal{B}_{(1-\kappa)\varpi}(m_1, m_2)|^p (d\kappa)^\varpi \right)^{\frac{1}{p}}$$

$$\times \left(\frac{\Gamma(1+\varpi)(|\Phi^{(\varpi)}(\vartheta_2)|^q - |\Phi^{(\varpi)}(\vartheta_1)|^q)}{\Gamma(1+2\varpi)} + \frac{|\Phi^{(\varpi)}(\vartheta_1)|^q}{\Gamma(1+\varpi)} \right)^{\frac{1}{q}}$$

which completes the proof of the Theorem 61. $\square$

Remark 3.9. *If we set $\varpi = 1$ in Theorem 3.25, then the following inequality holds:*

$$\left| \frac{\Phi(\vartheta_1) + \Phi(\vartheta_2)}{2} \mathcal{B}(m_1, m_2) - \frac{1}{2} \frac{1}{(\vartheta_2 - \vartheta_1)^{(m_1+m_2-1)}} \int_{\vartheta_1}^{\vartheta_2} \Omega(\varkappa)\Phi(\varkappa)d\varkappa \right|$$

$$\leq \frac{(\vartheta_2 - \vartheta_1)}{2} \left(\int_0^1 |\mathcal{B}_\kappa(m_1, m_2) - \mathcal{B}_{(1-\kappa)}(m_1, m_2)|^p d\kappa \right)^{\frac{1}{p}} \left(\frac{|\Phi'(\vartheta_2)|^q + |\Phi'(\vartheta_1)|^q}{2} \right)^{\frac{1}{q}} \tag{32}$$

which is proved by M. Z. Sarikaya in [26].
Here we have other scenarios as:

- *If in equation (62), we get $m_1 = 1$, $m_2 = \varpi$, (or $m_1 = \varpi$, $m_2 = 1$) then the inequality reduces to*

$$\left| \frac{\Phi(\vartheta_1) + \Phi(\vartheta_2)}{2} - \frac{\Gamma(\varpi+1)}{2(\vartheta_2 - \vartheta_1)^\varpi} \left[J^\varpi_{\vartheta_1+} \Phi(\vartheta_2) + J^\varpi_{\vartheta_2-} \Phi(\vartheta_1) \right] \right|$$

$$= \frac{(\vartheta_2 - \vartheta_1)}{(\varpi+1)} \left(1 - \frac{1}{2\varpi} \right) \left[\frac{|\Phi'(\vartheta_2)|^q + |\Phi'(\vartheta_1)|^q}{2} \right]^{\frac{1}{q}} \tag{33}$$

which is proved in [31].

- *If in equation (62), we get* $\mathfrak{m}_1 = 1$, $\mathfrak{m}_2 = \frac{\varpi}{k}$, *(or* $\mathfrak{m}_2 = 1$, $\mathfrak{m}_1 = \frac{\varpi}{k}$*) then the inequalities becomes*

$$\left| \frac{\Phi(\vartheta_1) + \Phi(\vartheta_2)}{2} - \frac{\Gamma_k(\varpi + k)}{2(\vartheta_2 - \vartheta_1)^{\frac{\varpi}{k}}} \left[J^{\varpi}_{\vartheta_1^+, k} \Phi(\vartheta_2) + J^{\varpi}_{\vartheta_2^-, k} \Phi(\vartheta_1) \right] \right|$$

$$= \frac{(\vartheta_2 - \vartheta_1)}{2(\frac{\varpi}{k}\mathfrak{p} + 1)^{\frac{1}{\mathfrak{p}}}} \left[\frac{|\Phi'(\vartheta_2)|^{\mathfrak{q}} + |\Phi'(\vartheta_1)|^{\mathfrak{q}}}{2} \right]^{\frac{1}{\mathfrak{q}}} \tag{34}$$

which is proved by Hussain et. al. in [28].

- *If in equation (62), we get* $\mathfrak{m}_1 = \mathfrak{m}_2 = 1$ *then the inequality reduces to Theorem (2.3) proved by S. S. Dragomir et al. in [29].*

3.2 Result's for Fractal Midpoint Inequalities

New Hermite-Hadamard type Lemma involving fractal Beta function is presented in this section.

Lemma 3.10. *Let* $I \subset \mathbb{R}$ *be an interval,* $\Phi : I^o \subset \mathbb{R} \to \mathbb{R}^{\varpi}$ *(* I^o *is the interior of I) such that* $\Phi \in D_{\varpi}(I^o)$ *and* $\Phi^{\varpi} \in C_{\varpi}[\vartheta_1, \vartheta_2]$ *for* $\vartheta_1, \vartheta_2 \in I^o$ *with* $\vartheta_1 < \vartheta_2$, *we have the identity:*

$$\Phi\left(\frac{\vartheta_1 + \vartheta_2}{2} \right) \beta_{\varpi}(\mathfrak{m}_1, \mathfrak{m}_2) - \frac{\Gamma(\varpi + 1)}{2^{\varpi}} \frac{1}{(\vartheta_2 - \vartheta_1)^{(\mathfrak{m}_1 + \mathfrak{m}_2 - 1)\varpi}} {}_{\vartheta_1}I^{(\varpi)}_{\vartheta_2} \Omega \Phi = \frac{(\vartheta_2 - \vartheta_1)^{\varpi}}{2^{\varpi}} \sum_{\ell=1}^{\ell=4} \kappa_{\ell} \tag{35}$$

where

$$\kappa_1 = \frac{1}{\Gamma(1 + \varpi)} \int_0^{\frac{1}{2}} [\beta_{\kappa\varpi}(\mathfrak{m}_1, \mathfrak{m}_2)] \Phi^{(\varpi)} (\kappa\vartheta_2 + (1 - \kappa)\vartheta_1) (d\kappa)^{\varpi},$$

$$\kappa_2 = \frac{1}{\Gamma(1 + \varpi)} \int_0^{\frac{1}{2}} [-\beta_{\kappa\varpi}(\mathfrak{m}_1, \mathfrak{m}_2)] \Phi^{(\varpi)} (\kappa\vartheta_1 + (1 - \kappa)\vartheta_2) (d\kappa)^{\varpi},$$

$$\kappa_3 = \frac{1}{\Gamma(1 + \varpi)} \int_{\frac{1}{2}}^1 [-\beta_{(1-\kappa)\varpi}(\mathfrak{m}_2, \mathfrak{m}_1)] \Phi^{(\varpi)} (\kappa\vartheta_2 + (1 - \kappa)\vartheta_1) (d\kappa)^{\varpi},$$

$$\kappa_4 = \frac{1}{\Gamma(1 + \varpi)} \int_{\frac{1}{2}}^1 [\beta_{(1-\kappa)\varpi}(\mathfrak{m}_2, \mathfrak{m}_1)] \Phi^{(\varpi)} (\kappa\vartheta_1 + (1 - \kappa)\vartheta_2) (d\kappa)^{\varpi},$$

for $\mathfrak{m}_1, \mathfrak{m}_2 > 0$.

16

Proof. Here using integration by parts, we have

$$\kappa_1 = \frac{1}{\Gamma(1+\varpi)} \int_0^{\frac{1}{2}} [\beta_{\kappa\varpi}(m_1,m_2)] \Phi^{(\varpi)} (\kappa\vartheta_2 + (1-\kappa)\vartheta_1) (d\kappa)^\varpi$$

$$= \frac{\beta_{\frac{\varpi}{2}}(m_1,m_2)}{(\vartheta_2-\vartheta_1)^\varpi} \Phi\left(\frac{\vartheta_1+\vartheta_2}{2}\right) - \frac{\Gamma(1+\varpi)}{(\vartheta_2-\vartheta_1)^\varpi}$$

$$\left(\frac{1}{\Gamma(1+\varpi)} \int_0^{\frac{1}{2}} \kappa^{(m_1-1)\varpi}(1-\kappa)^{(m_2-1)\varpi} \Phi(\kappa\vartheta_2 + (1-\kappa)\vartheta_1) (d\kappa)^\varpi \right),$$

$$\kappa_2 = \frac{1}{\Gamma(1+\varpi)} \int_0^{\frac{1}{2}} [-\beta_{\kappa\varpi}(m_1,m_2)] \Phi^{(\varpi)} (\kappa\vartheta_1 + (1-\kappa)\vartheta_2) (d\kappa)^\varpi$$

$$= \frac{\beta_{\frac{\varpi}{2}}(m_1,m_2)}{(\vartheta_2-\vartheta_1)^\varpi} \Phi\left(\frac{\vartheta_1+\vartheta_2}{2}\right) - \frac{\Gamma(1+\varpi)}{(\vartheta_2-\vartheta_1)^\varpi}$$

$$\left(\frac{1}{\Gamma(1+\varpi)} \int_0^{\frac{1}{2}} \kappa^{(m_1-1)\varpi}(1-\kappa)^{(m_2-1)\varpi} \Phi(\kappa\vartheta_1 + (1-\kappa)\vartheta_2) (d\kappa)^\varpi \right),$$

$$\kappa_3 = \frac{1}{\Gamma(1+\varpi)} \int_{\frac{1}{2}}^{1} [-\beta_{(1-\kappa)\varpi}(m_2,m_1)] \Phi^{(\varpi)} (\kappa\vartheta_2 + (1-\kappa)\vartheta_1) (d\kappa)^\varpi$$

$$= \frac{\beta_{\frac{\varpi}{2}}(m_2,m_1)}{(\vartheta_2-\vartheta_1)^\varpi} \Phi\left(\frac{\vartheta_1+\vartheta_2}{2}\right) - \frac{\Gamma(1+\varpi)}{(\vartheta_2-\vartheta_1)^\varpi}$$

$$\left(\frac{1}{\Gamma(1+\varpi)} \int_0^{\frac{1}{2}} \kappa^{(m_2-1)\varpi}(1-\kappa)^{(m_1-1)\varpi} \Phi(\kappa\vartheta_2 + (1-\kappa)\vartheta_1) (d\kappa)^\varpi \right),$$

$$\kappa_4 = \frac{1}{\Gamma(1+\varpi)} \int_{\frac{1}{2}}^{1} [\beta_{(1-\kappa)\varpi}(m_2,m_1)] \Phi^{(\varpi)} (\kappa\vartheta_1 + (1-\kappa)\vartheta_2) (d\kappa)^\varpi$$

$$= \frac{\beta_{\frac{\varpi}{2}}(m_2,m_1)}{(\vartheta_2-\vartheta_1)^\varpi} \Phi\left(\frac{\vartheta_1+\vartheta_2}{2}\right) - \frac{\Gamma(1+\varpi)}{(\vartheta_2-\vartheta_1)^\varpi}$$

$$\times \left(\frac{1}{\Gamma(1+\varpi)} \int_0^{\frac{1}{2}} \kappa^{(m_2-1)\varpi}(1-\kappa)^{(m_1-1)\varpi} \Phi(\kappa\vartheta_1 + (1-\kappa)\vartheta_2) (d\kappa)^\varpi \right),$$

Thus, by the above expression, the desired equality (66) is obtained. $\square$

Remark 3.11. *The identity (66) reduces to following result by alternating the variables*

$$\Phi\left(\frac{\vartheta_1+\vartheta_2}{2}\right) \beta_\varpi(m_1,m_2) - \frac{\Gamma(\varpi+1)}{2^\varpi} \frac{1}{(\vartheta_2-\vartheta_1)^{(m_1+m_2-1)\varpi}} {}_{\vartheta_1}I_{\vartheta_2}^{(\varpi)} \Omega\Phi$$

$$= \frac{(\vartheta_2-\vartheta_1)^\varpi}{2^\varpi} \frac{1}{\Gamma(1+\varpi)} \int_0^{\frac{1}{2}} \left[\beta_{\kappa\varpi}(m_1,m_2) + \beta_{\kappa\varpi}(m_2,m_1) \right] \tag{36}$$

$$\times \left[\Phi^{(\varpi)} (\kappa\vartheta_2 + (1-\kappa)\vartheta_1) - \Phi^{(\varpi)} (\kappa\vartheta_1 + (1-\kappa)\vartheta_2) \right] (d\kappa)^\varpi.$$

Theorem 3.12. *Let* $\Phi : [\vartheta_1, \vartheta_2] \to \mathbb{R}$ *be a differential mapping on* $(\vartheta_1, \vartheta_2)$ *with* $\vartheta_1 < \vartheta_2$. *If* $|\Phi^{(\varpi)}|$ *is convex on* $[\vartheta_1, \vartheta_2]$, *then the following equalities holds:*

$$\left| \Phi\left(\frac{\vartheta_1 + \vartheta_2}{2}\right) \beta_\varpi(\mathrm{m}_1, \mathrm{m}_2) - \frac{\Gamma(\varpi+1)}{2^\varpi} \frac{1}{(\vartheta_2 - \vartheta_1)^{(\mathrm{m}_1 + \mathrm{m}_2 - 1)\varpi}} \, _{\vartheta_1}I_{\vartheta_2}^{(\varpi)} \Omega\Phi \right|$$

$$\leq \frac{(\vartheta_2 - \vartheta_1)^\varpi}{2^\varpi} \left[|\Phi^{(\varpi)}(\vartheta_1)| + |\Phi^{(\varpi)}(\vartheta_2)| \right]$$

$$\times \left(\left(\frac{1}{2}\right)^\varpi \beta_\varpi(\mathrm{m}_1, \mathrm{m}_2) - \beta_{\frac{\varpi}{2}}(\mathrm{m}_1 + 1, \mathrm{m}_2) - \beta_{\frac{\varpi}{2}}(\mathrm{n} + 1, \mathrm{m}_1) \right) \tag{37}$$

for $\mathrm{m}_1, \mathrm{m}_2 > 0$.

Proof. By using Remark 3.11 with convexity of $|\Phi^{(\varpi)}|$, then we have

$$\left| \Phi\left(\frac{\vartheta_1 + \vartheta_2}{2}\right) \beta_\varpi(\mathrm{m}_1, \mathrm{m}_2) - \frac{\Gamma(\varpi+1)}{2^\varpi} \frac{1}{(\vartheta_2 - \vartheta_1)^{(\mathrm{m}_1 + \mathrm{m}_2 - 1)\varpi}} \, _{\vartheta_1}I_{\vartheta_2}^{(\varpi)} \Omega\Phi \right|$$

$$\leq \frac{(\vartheta_2 - \vartheta_1)^\varpi}{2^\varpi} \left[|\Phi^{(\varpi)}(\vartheta_1)| + |\Phi^{(\varpi)}(\vartheta_2)| \right]$$

$$\times \left(\frac{1}{\Gamma(1+\varpi)} \int_0^{\frac{1}{2}} \left[\beta_{\kappa\varpi}(\mathrm{m}_1, \mathrm{m}_2) + \beta_{\kappa\varpi}(\mathrm{m}_2, \mathrm{m}_1) \right] \right). \tag{38}$$

Alternating variable in the integrals, we get

$$\frac{1}{\Gamma(1+\varpi)} \int_0^{\frac{1}{2}} \left[\beta_{\kappa\varpi}(\mathrm{m}_1, \mathrm{m}_2) + \beta_{\kappa\varpi}(\mathrm{m}_2, \mathrm{m}_1) \right] (d\kappa)^\varpi$$

$$= \frac{1}{\Gamma(1+\varpi)} \int_0^{\frac{1}{2}} \int_0^\kappa \mathfrak{s}^{(\mathrm{m}_1 - 1)\varpi} (1 - \mathfrak{s})^{(\mathrm{m}_2 - 1)\varpi} (d\mathfrak{s})^\varpi (d\kappa)^\varpi$$

$$+ \frac{1}{\Gamma(1+\varpi)} \int_0^{\frac{1}{2}} \int_0^\kappa \mathfrak{s}^{(\mathrm{m}_2 - 1)\varpi} (1 - \mathfrak{s})^{(\mathrm{m}_1 - 1)\varpi} (d\mathfrak{s})^\varpi (d\kappa)^\varpi$$

$$= \left(\frac{1}{2}\right)^\varpi \beta_\varpi(\mathrm{m}_1, \mathrm{m}_2) - \beta_{\frac{\varpi}{2}}(\mathrm{m}_1 + 1, \mathrm{m}_2) - \beta_{\frac{\varpi}{2}}(\mathrm{n} + 1, \mathrm{m}_1). \tag{39}$$

By using equation (70) in the equation (69), we get the required result. $\qquad\square$

Remark 3.13. *If in Theorem 3.31, we get* $\varpi = 1$, *then, the inequality (3.31) becomes,*

$$\left| \Phi\left(\frac{\vartheta_1 + \vartheta_2}{2}\right) \beta(\mathrm{m}_1, \mathrm{m}_2) - \frac{1}{2(\vartheta_2 - \vartheta_1)^{(\mathrm{m}_1 + \mathrm{m}_2 - 1)}} \int_{\vartheta_1}^{\vartheta_2} \Omega(\varkappa)\Phi(\varkappa) d\varkappa \right|$$

$$= \frac{(\vartheta_2 - \vartheta_1)}{2} \left(\frac{1}{2}\beta(\mathrm{m}_1, \mathrm{m}_2) - \beta_{\frac{1}{2}}(\mathrm{m}_1 + 1, \mathrm{m}_2) - \beta_{\frac{1}{2}}(\mathrm{n} + 1, \mathrm{m}_1) \right) [|\Phi'(\vartheta_1)| + |\Phi'(\vartheta_2)|] \tag{40}$$

which is proved by M. Z. Sarikaya in [26].
Here we have other scenarios as:

- *If in equation (72), we get $\mathfrak{m}_1 = 1$, $\mathfrak{m}_2 = \varpi$, (or $\mathfrak{m}_1 = \varpi$, $\mathfrak{m}_2 = 1$) then the inequality reduces to*

$$\left| \frac{\Phi(\vartheta_1) + \Phi(\vartheta_2)}{2} - \frac{\Gamma(\varpi + 1)}{2(\vartheta_2 - \vartheta_1)^\varpi} \left[J^\varpi_{\vartheta_1^+} \Phi(\vartheta_2) + J^\varpi_{\vartheta_2, k} \Phi(\vartheta_1) \right] \right|$$
$$= \frac{(\vartheta_2 - \vartheta_1)}{2^{\varpi+1}(\varpi + 1)} \left[|\Phi'(\vartheta_2)| + |\Phi'(\vartheta_1)| \right]. \tag{41}$$

which is proved by M. Iqbal in [30].

- *If in equation (72), we get $\mathfrak{m}_1 = 1$, $\mathfrak{m}_2 = \frac{\varpi}{k}$, (or $\mathfrak{m}_2 = 1$, $\mathfrak{m}_1 = \frac{\varpi}{k}$) then the inequalities becomes*

$$\left| \frac{\Phi(\vartheta_1) + \Phi(\vartheta_2)}{2} - \frac{\Gamma_k(\varpi + k)}{2(\vartheta_2 - \vartheta_1)^{\frac{\varpi}{k}}} \left[J^\varpi_{\vartheta_1^+, k} \Phi(\vartheta_2) + J^\varpi_{\vartheta_2^-, k} \Phi(\vartheta_1) \right] \right|$$
$$= \frac{(\vartheta_2 - \vartheta_1)}{2^{\frac{\varpi}{k}+1}(\frac{\varpi}{k} + 1)} \left[|\Phi'(\vartheta_2)| + |\Phi'(\vartheta_1)| \right]. \tag{42}$$

which is given by Sarikaya and Ertugral in [32].

3.3 Result's for Fractal Trapezoidal Inequalities in Mercer sense

New Hermite-Hadamard mercer type Lemma involving Beta function in Yang's Calculus is presented in this section.:

Lemma 3.14. *Let $I \subset \mathbb{R}$ be an interval, $\Phi : I^o \subset \mathbb{R} \to \mathbb{R}^\varpi$ (I^o is the interior of I) such that $\Phi \in D_\varpi(I^o)$ and $\Phi^\varpi \in C_\varpi[\vartheta_1, \vartheta_2]$ for $\vartheta_1, \vartheta_2 \in I^o$ with $\vartheta_1 < \vartheta_2$ for all $\varkappa \in [\vartheta_1, \vartheta_2]$, we have the identity*

$$\frac{\Phi(\vartheta_1 + \vartheta_2 - \zeta_1) + \Phi(\vartheta_1 + \vartheta_2 - \zeta_2)}{2^\varpi} \beta_\varpi(\mathfrak{m}, \mathfrak{n}) - \frac{\Gamma(\varpi + 1)}{2^\varpi}$$

$$\times \frac{1}{(\zeta_2 - \zeta_1)^{(\mathfrak{m}+\mathfrak{n}-1)\varpi}} {}_{\vartheta_1 + \vartheta_2 - \zeta_2} I^{(\varpi)}_{\vartheta_1 + \vartheta_2 - \zeta_1} \Omega\Phi = \frac{(\zeta_2 - \zeta_1)^\varpi}{2^\varpi} \frac{1}{\Gamma(1 + \varpi)} \int_0^1 \beta_{\kappa\varpi}(\mathfrak{m}_1, \mathfrak{m}_2)$$

$$\left[\Phi^{(\varpi)}(\vartheta_1 + \vartheta_2 - (\kappa\zeta_2 + (1 - \kappa)\zeta_1)) - \Phi^{(\varpi)}(\vartheta_1 + \vartheta_2 - (\kappa\zeta_1 + (1 - \kappa)\zeta_2)) \right] (d\kappa)^\varpi$$

$$\tag{43}$$

where $\beta_{\kappa\varpi}(\mathfrak{m}_1, \mathfrak{m}_2)$ is incomplete beta function defined by

$$\beta_{\kappa\varpi}(\mathfrak{m}_1, \mathfrak{m}_2) = \int_0^\kappa \mathfrak{s}^{(\mathfrak{m}_1 - 1)\varpi}(1 - \mathfrak{s})^{(\mathfrak{m}_2 - 1)\varpi}(d\mathfrak{s})^\varpi, \qquad 0 \le \kappa \le 1$$

and

$$\Omega(\varkappa) = ((\vartheta_1 + \vartheta_2 - \zeta_1) - \varkappa)^{(\mathfrak{m}_1 - 1)\varpi}(\varkappa - (\vartheta_1 + \vartheta_2 - \zeta_2))^{(\mathfrak{m}_2 - 1)\varpi}$$

$$+ ((\vartheta_1 + \vartheta_2 - \zeta_1) - \varkappa)^{(\mathfrak{m}_2 - 1)\varpi}(\varkappa - (\vartheta_1 + \vartheta_2 - \zeta_2))^{(\mathfrak{m}_1 - 1)\varpi} \tag{44}$$

for $\mathfrak{m}_1, \mathfrak{m}_2 > 0$.

Proof. Here, we consider following integral and using integration by parts, then we have

$$\Phi_1 = \frac{1}{\Gamma(1 + \varpi)} \int_0^1 \beta_{t\varpi}(\mathfrak{m}_1, \mathfrak{m}_2) \Phi^{(\varpi)}(\vartheta_1 + \vartheta_2 - (\kappa\zeta_2 + (1 - \kappa)\zeta_1))(d\kappa)^\varpi$$

$$= -\frac{\beta_\varpi(\mathfrak{m}_1, \mathfrak{m}_2)\Phi(\vartheta_1 + \vartheta_2 - \zeta_2)}{(\zeta_2 - \zeta_1)^\varpi} + \frac{\Gamma(1 + \varpi)}{(\zeta_2 - \zeta_1)^\varpi}$$

$$\times \left(\frac{1}{\Gamma(1 + \varpi)} \int_0^1 \Phi(\vartheta_1 + \vartheta_2 - (\kappa\zeta_2 + (1 - \kappa)\zeta_1)) \kappa^{(\mathfrak{m}_1 - 1)\varpi}(1 - \kappa)^{(\mathfrak{m}_2 - 1)\varpi}(d\kappa)^\varpi \right)$$

$$= -\frac{\beta_\varpi(\mathfrak{m}_1, \mathfrak{m}_2)\Phi(\vartheta_1 + \vartheta_2 - \zeta_2)}{(\zeta_2 - \zeta_1)^\varpi} + \frac{\Gamma(1 + \varpi)}{(\zeta_2 - \zeta_1)^\varpi} \frac{1}{(\zeta_2 - \zeta_1)^{(\mathfrak{m}_1 + \mathfrak{m}_2 - 1)\varpi}} \left(\frac{1}{\Gamma(1 + \varpi)} \right.$$

$$\left. \times \int_{\vartheta_1 + \vartheta_2 - \zeta_2}^{\vartheta_1 + \vartheta_2 - \zeta_1} ((\vartheta_1 + \vartheta_2 - \zeta_1) - \varkappa)^{(\mathfrak{m}_1 - 1)\varpi}(\varkappa - (\vartheta_1 + \vartheta_2 - \zeta_2))^{(\mathfrak{m}_2 - 1)\varpi} \Phi(\varkappa)(d\varkappa)^\varpi \right).$$

And similarly, we obtain

$$
\Phi_2 = \frac{1}{\Gamma(1+\varpi)} \int_0^1 \beta_{t\varpi}(m_1, m_2) \Phi^{(\varpi)}(\vartheta_1 + \vartheta_2 - (\kappa \zeta_1 + (1-\kappa)\zeta_2))(d\kappa)^\varpi
$$

$$
= \frac{\beta_\varpi(m_1, m_2)\Phi(\vartheta_1 + \vartheta_2 - \zeta_1)}{(\zeta_2 - \zeta_1)^\varpi} - \frac{\Gamma(1+\varpi)}{(\zeta_2 - \zeta_1)^\varpi}
$$

$$
\left(\frac{1}{\Gamma(1+\varpi)} \int_0^1 \Phi(\vartheta_1 + \vartheta_2 - (\kappa \zeta_1 + (1-\kappa))\zeta_2) \, \kappa^{(m_1-1)\varpi}(1-\kappa)^{(m_2-1)\varpi}(d\kappa)^\varpi \right)
$$

$$
= \frac{\beta_\varpi(m_1, m_2)\Phi(\vartheta_1 + \vartheta_2 - \zeta_1)}{(\zeta_2 - \zeta_1)^\varpi} - \frac{\Gamma(1+\varpi)}{(\zeta_2 - \zeta_1)^\varpi} \frac{1}{(\zeta_2 - \zeta_1)^{(m_1+m_2-1)\varpi}} \left(\frac{1}{\Gamma(1+\varpi)} \right.
$$

$$
\left. \times \int_{\vartheta_1 + \vartheta_2 - \zeta_2}^{\vartheta_1 + \vartheta_2 - \zeta_1} (\varkappa - (\vartheta_1 + \vartheta_2 - \zeta_2))^{(m_1-1)\varpi} \left((\vartheta_1 + \vartheta_2 - \zeta_1) - \varkappa \right)^{(m_2-1)\varpi} \Phi(\varkappa)(d\varkappa)^\varpi \right).
$$

If we subtract Φ_1 from Φ_2 and multiply by $\frac{(\zeta_2 - \zeta_1)^\varpi}{2\varpi}$ then the proof is completed. $\qquad\square$

Remark 3.15. *Substituting $\vartheta_1 = \zeta_1$ and $\vartheta_2 = \zeta_2$ in Lemma 43, then the following equality holds:*

$$
\frac{\Phi(\zeta_1) + \Phi(\zeta_2)}{2\varpi} \beta_\varpi(m_1, m_2) - \frac{\Gamma(\varpi+1)}{2\varpi} \frac{1}{(\zeta_2 - \zeta_1)^{(m_1+m_2-1)\varpi}} \zeta_1 I_{\zeta_2}^{(\varpi)} \Omega\Phi
$$

$$
= \frac{(\zeta_2 - \zeta_1)^\varpi}{2\varpi} \frac{1}{\Gamma(1+\varpi)} \int_0^1 \beta_{\kappa\varpi}(m_1, m_2)
$$

$$
\left[\Phi^{(\varpi)}(\kappa \zeta_2 + (1-\kappa)\zeta_1) - \Phi^{(\varpi)}(\kappa \zeta_1 + (1-\kappa)\zeta_2) \right](d\kappa)^\varpi \tag{45}
$$

which is a new equality in literature.

Remark 3.16. *If in Lemma 43, we get $\vartheta_1 = \zeta_1$ and $\vartheta_2 = \zeta_2$, and $\varpi = 1$, then the following equality holds:*

$$
\frac{\Phi(\zeta_1) + \Phi(\zeta_2)}{2} \beta(m_1, m_2) - \frac{1}{2} \frac{1}{(\zeta_2 - \zeta_1)^{(m_1+m_2-1)}} \int_{\zeta_1}^{\zeta_2} \Omega(\varkappa)\Phi(\varkappa)d\varkappa
$$

$$
= \frac{(\zeta_2 - \zeta_1)}{2} \int_0^1 \beta_\kappa(m_1, m_2)[\Phi'(\kappa \zeta_2 + (1-\kappa)\zeta_1) - \Phi'(\kappa \zeta_1 + (1-\kappa)\zeta_2)]d\kappa \tag{46}
$$

which is proved by M. Z. Sarikaya in [26].
Here we have some scenarios as:
If in equation 46, we get $m_1 = 1$, $m_2 = \varpi$, (or $m_2 = 1$, $m_1 = \varpi$) then we have

$$
\frac{\Phi(\zeta_1) + \Phi(\zeta_2)}{2} - \frac{\Gamma(\varpi+1)}{(\zeta_2 - \zeta_1)^\varpi} \left[J_{\zeta_1^+}^\varpi \Phi(\zeta_2) + J_{\zeta_2^-}^\varpi \Phi(\zeta_1) \right]
$$

$$
= \frac{(\zeta_2 - \zeta_1)}{2} \int_0^1 [(1-\kappa)^\varpi - \kappa^\varpi]\Phi'(\kappa \zeta_1 + (1-\kappa)\zeta_2)d\kappa \tag{47}
$$

which is proved by Sarikaya et al. in [27].

If in equation 46, we get $m_1 = 1$, $m_2 = \frac{\varpi}{k}$, (or $m_2 = 1, m_1 = \frac{\varpi}{k}$) then we have

$$\frac{\Phi(\zeta_1) + \Phi(\zeta_2)}{2} - \frac{\Gamma_k(\varpi + k)}{(\zeta_2 - \zeta_1)^{\frac{\varpi}{k}}} \left[J^{\varpi}_{\zeta_1^+, k} \Phi(\zeta_2) + J^{\varpi}_{\zeta_2^-, k} \Phi(\zeta_1) \right]$$

$$= \frac{(\zeta_2 - \zeta_1)}{2} \int_0^1 [(1 - \kappa)^{\frac{\varpi}{k}} - \kappa^{\frac{\varpi}{k}}] \Phi'(\kappa \zeta_1 + (1 - \kappa)\zeta_2) d\kappa \tag{48}$$

which is proved by Hussain in [28].

If in Lemma 43, we get $\vartheta_1 = \zeta_1$ and $\vartheta_2 = \zeta_2$ and $\varpi = m_2 = m_1 = 1$, then the equality holds:

$$\frac{\Phi(\zeta_2) + \Phi(\zeta_1)}{2} - \left(\frac{1}{\zeta_2 - \zeta_1} \right) \int_{\zeta_1}^{\zeta_2} \Phi(\varkappa) d\varkappa$$

$$= \frac{(\zeta_2 - \zeta_1)}{2} \int_0^1 (1 - 2\kappa) \Phi'(\kappa \zeta_1 + (1 - \kappa)\zeta_2) d\kappa. \tag{49}$$

which is proved by Dragomir et. al. in [29].

Remark 3.17. *By the change of variable in Lemma 43, the identity 43 reduces to*

$$\frac{\Phi(\vartheta_1 + \vartheta_2 - \zeta_1) + \Phi(\vartheta_1 + \vartheta_2 - \zeta_2)}{2^{\varpi}} \beta_{\varpi}(m_1, m_2) - \frac{\Gamma(\varpi + 1)}{2^{\varpi}}$$

$$\times \frac{1}{(\zeta_2 - \zeta_1)^{(m_1 + m_2 - 1)\varpi}} {}_{\vartheta_1 + \vartheta_2 - \zeta_2} I^{(\varpi)}_{\vartheta_1 + \vartheta_2 - \zeta_1} \Omega \Phi$$

$$= \frac{(\zeta_2 - \zeta_1)^{\varpi}}{2^{\varpi}} \frac{1}{\Gamma(1 + \varpi)} \int_0^1 \Phi^{\varpi} (\vartheta_1 + \vartheta_2 - (\kappa \zeta_1 + (1 - \kappa)\zeta_2))$$

$$\times [\beta_{\kappa\varpi}(m_1, m_2) - \beta_{(1 - \kappa)\varpi}(m_1, m_2)](d\kappa)^{\varpi}. \tag{50}$$

Now, we extend some estimates of the right hand side of a Hermite-Hadamard mercer type inequality for functions whose first derivatives absolute values are convex as follows:

Theorem 3.18. *Let Φ be defined as in Lemma 3.14 and if the mapping $|\Phi^{(\varpi)}|$ is convex on $[\vartheta_1, \vartheta_2]$, then :*

$$\left| \frac{\Phi(\vartheta_1 + \vartheta_2 - \zeta_1) + \Phi(\vartheta_1 + \vartheta_2 - \zeta_2)}{2^{\varpi}} \beta_{\varpi}(m_1, m_2) - \frac{\Gamma(\varpi + 1)}{2^{\varpi}} \right.$$

$$\left. \times \frac{1}{(\zeta_2 - \zeta_1)^{(m_1 + m_2 - 1)\varpi}} {}_{\vartheta_1 + \vartheta_2 - \zeta_2} I^{\varpi}_{\vartheta_1 + \vartheta_2 - \zeta_1} \Omega \Phi \right|$$

$$\leq \frac{(\zeta_2 - \zeta_1)^{\varpi}}{2^{\varpi}} \left[|\Phi^{(\varpi)}(\vartheta_1 + \vartheta_2 - \zeta_1)| + |\Phi^{(\varpi)}(\vartheta_1 + \vartheta_2 - \zeta_2)| \right]$$

$$\left(\frac{1}{\Gamma(1 + \varpi)} \int_0^{\frac{1}{2}} |\beta_{\kappa\varpi}(m_1, m_2) - \beta_{(1 - \kappa)\varpi}(m_1, m_2)|(d\kappa)^{\varpi} \right) \tag{51}$$

for $m_1, m_2 > 0$.

Proof. Utilizing Lemma 43 with the convexity of $|\Phi^{(\varpi)}|$ then we have

$$\left| \frac{\Phi(\vartheta_1 + \vartheta_2 - \zeta_1) + \Phi(\vartheta_1 + \vartheta_2 - \zeta_2)}{2^{\varpi}} \beta_{\varpi}(m_1, m_2) - \frac{\Gamma(\varpi + 1)}{2^{\varpi}} \right.$$

$$\left. \times \frac{1}{(\zeta_2 - \zeta_1)^{(m_1 + m_2 - 1)\varpi}} {}_{\vartheta_1 + \vartheta_2 - \zeta_2} I^{\varpi}_{\vartheta_1 + \vartheta_2 - \zeta_1} \Omega \Phi \right|$$

$$\leq \frac{(\zeta_2 - \zeta_1)^{\varpi}}{2^{\varpi}} \frac{1}{\Gamma(1 + \varpi)} \int_0^1 |\beta_{\kappa\varpi}(m_1, m_2) - \beta_{(1-\kappa)\varpi}(m_1, m_2)|$$

$$\times |\Phi^{(\varpi)}(\vartheta_1 + \vartheta_2 - (\kappa\zeta_1 + (1 - \kappa)\zeta_2|(d\kappa)^{\varpi})$$

$$\leq \frac{(\zeta_2 - \zeta_1)^{\varpi}}{2^{\varpi}} \left(\frac{1}{\Gamma(1 + \varpi)} \int_0^{\frac{1}{2}} [\beta_{\kappa\varpi}(m_1, m_2) - \beta_{(1-\kappa)\varpi}(m_1, m_2)] \left[\kappa^{\varpi} |\Phi^{(\varpi)}(\vartheta_1 + \vartheta_2 - \zeta_1)| \right. \right.$$

$$\left. + (1 - \kappa)^{\varpi} |\Phi^{(\varpi)}(\vartheta_1 + \vartheta_2 - \zeta_2)| \right] (d\kappa)^{\varpi} \right) + \frac{(\zeta_2 - \zeta_1)^{\varpi}}{2^{\varpi}} \left(\frac{1}{\Gamma(1 + \varpi)} \int_{\frac{1}{2}}^1 [\beta_{(1-\kappa)\varpi}(m_1, m_2) \right.$$

$$\left. - \beta_{\kappa\varpi}(m_1, m_2)] \left[\kappa^{\varpi} |\Phi^{(\varpi)}(\vartheta_1 + \vartheta_2 - \zeta_1)| + (1 - \kappa)^{\varpi} |\Phi^{(\varpi)}(\vartheta_1 + \vartheta_2 - \zeta_2)| \right] (d\kappa)^{\varpi} \right)$$

$$\leq \frac{(\zeta_2 - \zeta_1)^{\varpi}}{2^{\varpi}} [|\Phi^{(\varpi)}(\vartheta_1 + \vartheta_2 - \zeta_1)| + |\Phi^{(\varpi)}(\vartheta_1 + \vartheta_2 - \zeta_2)|]$$

$$\times \frac{1}{\Gamma(1 + \varpi)} \int_0^{\frac{1}{2}} |\beta_{\kappa\varpi}(m_1, m_2) - \beta_{(1-\kappa)\varpi}(m_1, m_2)|(d\kappa)^{\varpi}$$

which concludes the proof. $\qquad\square$

Remark 3.19. *If we set $\vartheta_1 = \zeta_1$, $\vartheta_2 = \zeta_2$ in Theorem 3.18, then following inequality holds:*

$$\left| \frac{\Phi(\zeta_1) + \Phi(\zeta_2)}{2^{\varpi}} \beta_{\varpi}(m_1, m_2) - \frac{\Gamma(\varpi + 1)}{2^{\varpi}} \frac{1}{(\zeta_2 - \zeta_1)^{(m_1 + m_2 - 1)\varpi}} {}_{\zeta_1} I^{\varpi}_{\zeta_2} \Omega \Phi \right|$$

$$\leq \frac{(\zeta_2 - \zeta_1)^{\varpi}}{2^{\varpi}} \left[|\Phi^{(\varpi)}(\zeta_1)| + |\Phi^{(\varpi)}(\zeta_2)| \right] \left(\frac{1}{\Gamma(1 + \varpi)} \right.$$

$$\left. \times \int_0^{\frac{1}{2}} |\beta_{\kappa\varpi}(m_1, m_2) - \beta_{(1-\kappa)\varpi}(m_1, m_2)|(d\kappa)^{\varpi} \right) \qquad (52)$$

which is a new inequality in literature.

Remark 3.20. *If in Theorem 3.18, we get $\vartheta_1 = \zeta_1$, $\vartheta_2 = \zeta_2$ and $\varpi = 1$, then the following inequality holds:,*

$$\left| \frac{\Phi(\zeta_1) + \Phi(\zeta_2)}{2} \beta(m_1, m_2) - \frac{1}{2} \frac{1}{(\zeta_2 - \zeta_1)^{(m_1 + m_2 - 1)}} \int_{\zeta_1}^{\zeta_2} \Omega(\varkappa) \Phi(\varkappa) d(\varkappa) \right|$$

$$\leq \frac{(\zeta_2 - \zeta_1)}{2} [|\Phi'(\zeta_1)| + |\Phi'(\zeta_2)|] \int_0^{\frac{1}{2}} |\beta_{\kappa}(m_1, m_2) - \beta_{(1-\kappa)}(m_1, m_2)| d\kappa, \qquad (53)$$

which is proved by M. Z. Sarikaya in [26].
Here we have other scenarios as:

If in equation 53, we get $\mathfrak{m}_1 = 1, \mathfrak{m}_2 = \varpi, (or\ \mathfrak{m}_2 = 1,\ \mathfrak{m}_1 = \varpi)$, *then we have*

$$\left| \frac{\Phi(\zeta_1) + \Phi(\zeta_2)}{2} - \frac{\Gamma(\varpi+1)}{(\zeta_2-\zeta_1)^\varpi} \left[J_{\zeta_1^+}^\varpi \Phi(\zeta_2) + J_{\zeta_2^-}^\varpi \Phi(\zeta_1) \right] \right|$$
$$= \frac{(\zeta_2-\zeta_1)}{2(\varpi+1)} \left(1 - \frac{1}{2^\varpi} \right) [\Phi'(\zeta_2) + \Phi'(\zeta_1)], \tag{54}$$

which is proved by M. Iqbal in [30].

If in equation 53, we get $\mathfrak{m}_1 = 1,\ \mathfrak{m}_2 = \frac{\varpi}{k}, (or\ \mathfrak{m}_2 = 1, \mathfrak{m}_1 = \frac{\varpi}{k})$ *then we have*

$$\left| \frac{\Phi(\zeta_1) + \Phi(\zeta_2)}{2} - \frac{\Gamma_k(\varpi+k)}{2(\zeta_2-\zeta_1)^\varpi} \left[J_{\zeta_1^+,k}^\varpi \Phi(\zeta_2) + J_{\zeta_2^-,k}^\varpi \Phi(\zeta_1) \right] \right|$$
$$= \frac{(\zeta_2-\zeta_1)}{(\frac{\varpi}{k}+1)} \left(1 - \frac{1}{2^{\frac{\varpi}{k}}} \right) [\Phi'(\zeta_2) + \Phi'(\zeta_1)], \tag{55}$$

which is proved by Hussain et. al. in [28].

Remark 3.21. *If in Theorem* 3.18, *we get* $\vartheta_1 = \zeta_1$, $\vartheta_2 = \zeta_2$ *and* $\varpi = \mathfrak{m}_1 = \mathfrak{m}_2 = 1$, *then, the inequality* 51 *becomes the inequality*

$$\left| \frac{\Phi(\zeta_1) + \Phi(\zeta_2)}{2} - \left(\frac{1}{\zeta_2-\zeta_1} \right) \int_{\zeta_1}^{\zeta_2} \Phi(\varkappa) d(\varkappa) \right|$$
$$\leq \frac{(\zeta_2-\zeta_1)}{8} [|\prime(\zeta_1)| + |\prime(\zeta_2)|], \tag{56}$$

which is proved by Dragomir et. al. in [29].

Theorem 3.22. *Let* Φ *be defined as in Lemma* 3.14 *and if the mapping* $|\Phi^{(\varpi)}|^q$, $q > 1$ *is convex on* $[\vartheta_1, \vartheta_2]$, *then we have the following inequality:*

$$\left| \frac{\Phi(\vartheta_1 + \vartheta_2 - \zeta_1) + \Phi(\vartheta_1 + \vartheta_2 - \zeta_2)}{2^\varpi} \beta_\varpi(\mathfrak{m}_1, \mathfrak{m}_2) - \frac{\Gamma(\varpi+1)}{2^\varpi} \right.$$
$$\left. \times \frac{1}{(\zeta_2-\zeta_1)^{(\mathfrak{m}_1+\mathfrak{m}_2-1)\varpi}} {}_{\vartheta_1+\vartheta_2-\zeta_2} I_{\vartheta_1+\vartheta_2-\zeta_1}^\varpi \Omega\Phi \right|$$
$$\leq \frac{(\zeta_2-\zeta_1)^\varpi}{2^\varpi} \left[\frac{1}{\mathfrak{p}} \left(\frac{1}{\Gamma(1+\varpi)} \int_0^1 |\beta_{\kappa\varpi}(\mathfrak{m}_1, \mathfrak{m}_2) - \beta_{(1-\kappa)\varpi}(\mathfrak{m}_1, \mathfrak{m}_2)|^\mathfrak{p} (d\kappa)^\varpi \right) \right.$$
$$\left. + \frac{1}{\mathfrak{q}} \left(\frac{\Gamma(1+\varpi)(|\Phi^{(\varpi)}(\vartheta_1 + \vartheta_2 - \zeta_1)|^\mathfrak{q} - |\Phi^{(\varpi)}(\vartheta_1 + \vartheta_2 - \zeta_2)|^\mathfrak{q})}{\Gamma(1+2\varpi)} + \frac{|\Phi^{(\varpi)}(\vartheta_1 + \vartheta_2 - \zeta_2)|}{\Gamma(\varpi+1)} \right) \right]$$
$$\tag{57}$$

for $\mathfrak{m}_1, \mathfrak{m}_2 > 0$, *where* $\frac{1}{\mathfrak{p}} + \frac{1}{\mathfrak{q}} = 1$.

Proof. Utilizing lemma 43 and *Young's* inequality with the convexity of $|\Phi^{(\varpi)}|^\mathfrak{q}$, we find that

$$\left| \frac{\Phi(\vartheta_1 + \vartheta_2 - \zeta_1) + \Phi(\vartheta_1 + \vartheta_2 - \zeta_2)}{2^\varpi} \beta_\varpi(m_1, m_2) - \frac{\Gamma(\varpi + 1)}{2^\varpi} \right.$$

$$\left. \times \frac{1}{(\zeta_2 - \zeta_1)^{(m_1 + m_2 - 1)\varpi}} \, {}_{\vartheta_1 + \vartheta_2 - \zeta_2} I^\varpi_{\vartheta_1 + \vartheta_2 - \zeta_1} \Omega\Phi \right|$$

$$\leq \frac{(\zeta_2 - \zeta_1)^\varpi}{2^\varpi} \left[\frac{1}{p} \left(\frac{1}{\Gamma(1 + \varpi)} \int_0^1 |\beta_{\kappa\varpi}(m_1, m_2) - \beta_{(1-\kappa)\varpi}(m_1, m_2)|^p (d\kappa)^\varpi \right) \right.$$

$$\left. + \frac{1}{q} \left(\frac{1}{\Gamma(1 + \varpi)} \int_0^1 \left| \Phi^{(\varpi)}(\vartheta_1 + \vartheta_2 - (\kappa\zeta_1 + (1 - \kappa)\zeta_2)) \right|^q (d\kappa)^\varpi \right) \right]$$

$$\leq \frac{(\zeta_2 - \zeta_1)^\varpi}{2^\varpi} \left[\frac{1}{p} \left(\frac{1}{\Gamma(1 + \varpi)} \int_0^1 |\beta_{\kappa\varpi}(m_1, m_2) - \beta_{(1-\kappa)\varpi}(m_1, m_2)|^p (d\kappa)^\varpi \right) \right.$$

$$\left. + \frac{1}{q} \left(\frac{1}{\Gamma(1 + \varpi)} \int_0^1 \left(\kappa^\varpi |\Phi^{(\varpi)}(\vartheta_1 + \vartheta_2 - \zeta_1)|^q + (1 - \kappa)^\varpi |\Phi^{(\varpi)}(\vartheta_1 + \vartheta_2 - \zeta_2)|^q \right) (d\kappa)^\varpi \right) \right]$$

$$\leq \frac{(\zeta_2 - \zeta_1)^\varpi}{2^\varpi} \left[\frac{1}{p} \left(\frac{1}{\Gamma(1 + \varpi)} \int_0^1 |\beta_{\kappa\varpi}(m_1, m_2) - \beta_{(1-\kappa)\varpi}(m_1, m_2)|^p (d\kappa)^\varpi \right) \right.$$

$$\left. + \frac{1}{q} \left(\frac{\Gamma(1 + \varpi)(|\Phi^{(\varpi)}(\vartheta_1 + \vartheta_2 - \zeta_1)|^q - |\Phi^{(\varpi)}(\vartheta_1 + \vartheta_2 - \zeta_2)|^q)}{\Gamma(1 + 2\varpi)} + \frac{|\Phi^{(\varpi)}(\vartheta_1 + \vartheta_2 - \zeta_2)|}{\Gamma(1 + \varpi)} \right) \right]$$

which completes the proof of the 61. $\qquad\square$

Remark 3.23. *If we set $\vartheta_1 = \zeta_1$, $\vartheta_2 = \zeta_2$ in Theorem 3.22, then following inequality holds:*

$$\left| \frac{\Phi(\zeta_1) + \Phi(\zeta_2)}{2^\varpi} \beta_\varpi(m_1, m_2) - \frac{\Gamma(\varpi + 1)}{2^\varpi} \frac{1}{(\zeta_2 - \zeta_1)^{(m_1 + m_2 - 1)\varpi}} \, {}_{\zeta_1} I^\varpi_{\zeta_2} \Omega\Phi \right|$$

$$\leq \frac{(\zeta_2 - \zeta_1)^\varpi}{2^\varpi} \left[\frac{1}{p} \left(\frac{1}{\Gamma(1 + \varpi)} \int_0^1 |\beta_{\kappa\varpi}(m_1, m_2) - \beta_{(1-\kappa)\varpi}(m_1, m_2)|^p (d\kappa)^\varpi \right) \right.$$

$$\left. + \frac{1}{q} \left(\frac{\Gamma(1 + \varpi)(|\Phi^{(\varpi)}(\zeta_2)|^q - |\Phi^{(\varpi)}(\zeta_1)|^q)}{\Gamma(1 + 2\varpi)} + \frac{|\Phi^{(\varpi)}(\zeta_1)|}{\Gamma(\varpi + 1)} \right) \right] \qquad (58)$$

which is a new inequality in literature.

Remark 3.24. *If in Theorem 3.22 we get $\vartheta_1 = \zeta_1$, $\vartheta_2 = \zeta_2$ and $\varpi = 1$, then the following inequality holds:*

$$\left| \frac{\Phi(\zeta_1) + \Phi(\zeta_2)}{2} \beta(m_1, m_2) - \frac{1}{2} \frac{1}{(\zeta_2 - \zeta_1)^{(m_1 + m_2 - 1)}} \int_{\zeta_1}^{\zeta_2} \Omega(\varkappa)\Phi(\varkappa) d\varkappa \right|$$

$$\leq \frac{(\zeta_2 - \zeta_1)}{2} \left[\frac{1}{p} \left(\int_0^1 |\beta_\kappa(m_1, m_2) - \beta_{(1-\kappa)}(m_1, m_2)|^p (d\kappa) \right) + \frac{1}{q} \left(\frac{|\Phi'(\zeta_2)|^q + |\Phi'(\zeta_1)|^q}{2} \right) \right] \qquad (59)$$

which is a new inequality in literature.

Theorem 3.25. *Let Φ be defined as in Lemma 3.14 and if the mapping $|\Phi^{(\varpi)}|^q$, $q > 1$ is convex on $[\vartheta_1, \vartheta_2]$, then we have the following inequality:*

$$\left| \frac{\Phi(\vartheta_1 + \vartheta_2 - \zeta_1) + \Phi(\vartheta_1 + \vartheta_2 - \zeta_2)}{2^\varpi} \beta_\varpi(m_1, m_2) - \frac{\Gamma(\varpi + 1)}{2^\varpi} \right.$$
$$\left. \times \frac{1}{(\zeta_2 - \zeta_1)^{(m_1 + m_2 - 1)\varpi}} {}_{\vartheta_1 + \vartheta_2 - \zeta_2} I^\varpi_{\vartheta_1 + \vartheta_2 - \zeta_1} \Omega\Phi \right|$$
$$\leq \frac{(\zeta_2 - \zeta_1)^\varpi}{2^\varpi} \left(\frac{1}{\Gamma(1 + \varpi)} \int_0^1 |\beta_{\kappa\varpi}(m_1, m_2) - \beta_{(1-\kappa)\varpi}(m_1, m_2)|^p (d\kappa)^\varpi \right)^{\frac{1}{p}}$$
$$\times \left(\frac{\Gamma(1 + \varpi)(|\Phi^{(\varpi)}(\vartheta_1 + \vartheta_2 - \zeta_1)|^q - |\Phi^{(\varpi)}(\vartheta_1 + \vartheta_2 - \zeta_2)|^q)}{\Gamma(1 + 2\varpi)} + \frac{|\Phi^{(\varpi)}(\vartheta_1 + \vartheta_2 - \zeta_2)|}{\Gamma(\varpi + 1)} \right)^{\frac{1}{q}}$$

$$(60)$$

for $m_1, m_2 > 0$, where $\frac{1}{p} + \frac{1}{q} = 1$.

Proof. Utilizing *Hölder's* inequality in Lemma 43 with the convexity of $|\Phi^{(\varpi)}|^q$ then we have

$$\left| \frac{\Phi(\vartheta_1 + \vartheta_2 - \zeta_1) + \Phi(\vartheta_1 + \vartheta_2 - \zeta_2)}{2^\varpi} \beta_\varpi(m_1, m_2) - \frac{\Gamma(\varpi + 1)}{2^\varpi} \right.$$
$$\left. \times \frac{1}{(\zeta_2 - \zeta_1)^{(m_1 + m_2 - 1)\varpi}} {}_{\vartheta_1 + \vartheta_2 - \zeta_2} I^\varpi_{\vartheta_1 + \vartheta_2 - \zeta_1} \Omega\Phi \right|$$
$$\leq \frac{(\zeta_2 - \zeta_1)^\varpi}{2^\varpi} \left(\frac{1}{\Gamma(1 + \varpi)} \int_0^1 |\beta_{\kappa\varpi}(m_1, m_2) - \beta_{(1-\kappa)\varpi}(m_1, m_2)|^p (d\kappa)^\varpi \right)^{\frac{1}{p}}$$
$$\times \left(\frac{1}{\Gamma(1 + \varpi)} \int_0^1 \left| \Phi^{(\varpi)}(\vartheta_1 + \vartheta_2 - (\kappa\zeta_1 + (1 - \kappa)\zeta_2)) \right|^q (d\kappa)^\varpi \right)^{\frac{1}{q}}$$
$$\leq \frac{(\zeta_2 - \zeta_1)^\varpi}{2^\varpi} \left(\frac{1}{\Gamma(1 + \varpi)} \int_0^1 |\beta_{\kappa\varpi}(m_1, m_2) - \beta_{(1-\kappa)\varpi}(m_1, m_2)|^p (d\kappa)^\varpi \right)^{\frac{1}{p}}$$
$$\times \left(\frac{1}{\Gamma(1 + \varpi)} \int_0^1 \left(\kappa^\varpi |\Phi^{(\varpi)}(\vartheta_1 + \vartheta_2 - \zeta_1)|^q + (1 - \kappa)^\varpi |\Phi^{(\varpi)}(\vartheta_1 + \vartheta_2 - \zeta_2)|^q \right) (d\kappa)^\varpi \right)^{\frac{1}{q}}$$
$$\leq \frac{(\zeta_2 - \zeta_1)^\varpi}{2^\varpi} \left(\frac{1}{\Gamma(1 + \varpi)} \int_0^1 |\beta_{\kappa\varpi}(m_1, m_2) - \beta_{(1-\kappa)\varpi}(m_1, m_2)|^p (d\kappa)^\varpi \right)^{\frac{1}{p}}$$
$$\left(\frac{\Gamma(1 + \varpi)(|\Phi^{(\varpi)}(\vartheta_1 + \vartheta_2 - \zeta_1)|^q - |\Phi^{(\varpi)}(\vartheta_1 + \vartheta_2 - \zeta_2)|^q)}{\Gamma(1 + 2\varpi)} + \frac{|\Phi^{(\varpi)}(\vartheta_1 + \vartheta_2 - \zeta_2)|}{\Gamma(1 + \varpi)} \right)^{\frac{1}{q}}$$

which completes the proof of the 61. $\qquad\square$

Remark 3.26. *If we set* $\vartheta_1 = \zeta_1$, $\vartheta_2 = \zeta_2$ *in Theorem 3.25, then following inequality holds:*

$$\left| \frac{\Phi(\zeta_1) + \Phi(\zeta_2)}{2^{\varpi}} \beta_{\varpi}(m_1, m_2) - \frac{\Gamma(\varpi+1)}{2^{\varpi}} \frac{1}{(\zeta_2 - \zeta_1)^{(m_1+m_2-1)\varpi}} \zeta_1 I_{\zeta_2}^{\varpi} \Omega \Phi \right|$$

$$\leq \frac{(\zeta_2 - \zeta_1)^{\varpi}}{2^{\varpi}} \left(\frac{1}{\Gamma(1+\varpi)} \int_0^1 |\beta_{\kappa\varpi}(m_1, m_2) - \beta_{(1-\kappa)\varpi}(m_1, m_2)|^p (d\kappa)^{\varpi} \right)^{\frac{1}{p}}$$

$$\times \left(\frac{\Gamma(1+\varpi)(|\Phi^{(\varpi)}(\zeta_2)|^q - |\Phi^{(\varpi)}(\zeta_1)|^q)}{\Gamma(1+2\varpi)} + \frac{|\Phi^{(\varpi)}(\zeta_1)|}{\Gamma(\varpi+1)} \right)^{\frac{1}{q}} \tag{61}$$

which is a new inequality in literature.

Remark 3.27. *If in Theorem 3.25 we get* $\vartheta_1 = \zeta_1$, $\vartheta_2 = \zeta_2$ *and* $\varpi = 1$, *then the following inequality holds:*

$$\left| \frac{\Phi(\zeta_1) + \Phi(\zeta_2)}{2} \beta(m_1, m_2) - \frac{1}{2} \frac{1}{(\zeta_2 - \zeta_1)^{(m_1+m_2-1)}} \int_{\zeta_1}^{\zeta_2} \Omega(\varkappa) \Phi(\varkappa) d\varkappa \right|$$

$$\leq \frac{(\zeta_2 - \zeta_1)}{2} \left(\int_0^1 |\beta_{\kappa}(m_1, m_2) - \beta_{(1-\kappa)}(m_1, m_2)|^p (d\kappa) \right)^{\frac{1}{p}} \left(\frac{|\Phi'(\zeta_2)|^q + |\Phi'(\zeta_1)|^q}{2} \right)^{\frac{1}{q}} \tag{62}$$

which is proved by M. Z. Sarikaya in [26]. Here we have some scenarios: If in equation 62, we get $m_1 = 1$, $m_2 = \varpi$, *(or* $m_1 = \varpi, m_2 = 1$*) then the inequality reduces to*

$$\left| \frac{\Phi(\zeta_1) + \Phi(\zeta_2)}{2} - \frac{\Gamma(\varpi+1)}{2(\zeta_2 - \zeta_1)^{\varpi}} \left[J_{\zeta_1^+}^{\varpi} \Phi(\zeta_2) + J_{\zeta_2, k}^{\varpi} \Phi(\zeta_1) \right] \right|$$

$$= \frac{(\zeta_2 - \zeta_1)}{(\varpi+1)} \left(1 - \frac{1}{2^{\varpi}} \right) \left[\frac{|\Phi'(\zeta_2)|^q + |\Phi'(\zeta_1)|^q}{2} \right]^{\frac{1}{q}} \tag{63}$$

which is proved in [31].

If in equation 62, we get $m_1 = 1$, $m_2 = \frac{\varpi}{k}$, *(or* $m_2 = 1, m_1 = \frac{\varpi}{k}$*) then the inequalities becomes*

$$\left| \frac{\Phi(\zeta_1) + \Phi(\zeta_2)}{2} - \frac{\Gamma_k(\varpi+k)}{2(\zeta_2 - \zeta_1)^{\frac{\varpi}{k}}} \left[J_{\zeta_1^+, k}^{\varpi} \Phi(\zeta_2) + J_{\zeta_2^-, k}^{\varpi} \Phi(\zeta_1) \right] \right|$$

$$= \frac{(\zeta_2 - \zeta_1)}{2(\frac{\varpi}{k}p+1)^{\frac{1}{p}}} \left[\frac{|\Phi'(\zeta_2)|^q + |\Phi'(\zeta_1)|^q}{2} \right]^{\frac{1}{q}} \tag{64}$$

which is proved by Hussain et. al. in [28].

Remark 3.28. *If in equation 62, we get* $\vartheta_1 = \zeta_1$, $\vartheta_2 = \zeta_2$ *and* $m_1 = m_2 = 1$ *then the inequality reduces to*

$$\left| \frac{\Phi(\zeta_1) + \Phi(\zeta_2)}{2} - \frac{1}{(\zeta_2 - \zeta_1)} \int_{\zeta_1}^{\zeta_2} \Phi(\varkappa) d\varkappa \right| \leq \frac{(\zeta_2 - \zeta_1)}{2(p+1)^{\frac{1}{p}}} \left[\frac{|\Phi'(\zeta_1)|^{\frac{p}{(p-1)}} + |\Phi'(\zeta_2)|^{\frac{p}{(p-1)}}}{2} \right]^{\frac{(p-1)}{p}} \tag{65}$$

which is proved in [29].

3.4 Result's for Fractal Midpoint Inequalities in Mercer sense

New Hermite-Hadamard mercer type Lemma involving Beta function in Yang's calculus is presented in this section.

Lemma 3.29. *Let $I \subset \mathbb{R}$ be an interval, $\Phi : I^o \subset \mathbb{R} \to \mathbb{R}^\varpi$ (I^o is the interior of I) such that $\Phi \in D_\varpi(I^o)$ and $\Phi^\varpi \in C_\varpi[\vartheta_1, \vartheta_2]$ for $\vartheta_1, \vartheta_2 \in I^o$ with $\vartheta_1 < \vartheta_2$ for all $\varkappa \in [\vartheta_1, \vartheta_2]$, we have the identity*

$$\Phi\left(\vartheta_1 + \vartheta_2 - \frac{\zeta_1 + \zeta_2}{2}\right)\beta_\varpi(m_1, m_2) - \frac{\Gamma(\varpi + 1)}{2^\varpi}\frac{1}{(\zeta_2 - \zeta_1)^{(m_1 + m_2 - 1)\varpi}}$$

$$\times {}_{\vartheta_1 + \vartheta_2 - \zeta_2}I^{(\varpi)}_{\vartheta_1 + \vartheta_2 - \zeta_1}\Omega\Phi = \frac{(\zeta_2 - \zeta_1)^\varpi}{2^\varpi}\sum_{\ell=1}^{\ell=4}\kappa_\ell \tag{66}$$

where

$$\kappa_1 = \frac{1}{\Gamma(1 + \varpi)}\int_0^{\frac{1}{2}}[-\beta_{\kappa\varpi}(m_1, m_2)]\Phi^{(\varpi)}(\vartheta_1 + \vartheta_2 - (\kappa\zeta_2 + (1 - \kappa)\zeta_1))(d\kappa)^\varpi,$$

$$\kappa_2 = \frac{1}{\Gamma(1 + \varpi)}\int_0^{\frac{1}{2}}[\beta_{\kappa\varpi}(m_1, m_2)]\Phi^{(\varpi)}(\vartheta_1 + \vartheta_2 - (\kappa\zeta_1 + (1 - \kappa)\zeta_2))(d\kappa)^\varpi,$$

$$\kappa_3 = \frac{1}{\Gamma(1 + \varpi)}\int_{\frac{1}{2}}^1[\beta_{(1-\kappa)\varpi}(m_2, m_1)]\Phi^{(\varpi)}(\vartheta_1 + \vartheta_2 - (\kappa\zeta_2 + (1 - \kappa)\zeta_1))(d\kappa)^\varpi,$$

$$\kappa_4 = \frac{1}{\Gamma(1 + \varpi)}\int_{\frac{1}{2}}^1[-\beta_{(1-\kappa)\varpi}(m_2, m_1)]\Phi^{(\varpi)}(\vartheta_1 + \vartheta_2 - (\kappa\zeta_1 + (1 - \kappa)\zeta_2))(d\kappa)^\varpi,$$

for $m_1, m_2 > 0$.

Proof. Here using integration by parts then we have

$$\kappa_1 = \frac{1}{\Gamma(1 + \varpi)}\int_0^{\frac{1}{2}}[-\beta_{\kappa\varpi}(m_1, m_2)]\Phi^{(\varpi)}(\vartheta_1 + \vartheta_2 - (\kappa\zeta_2 + (1 - \kappa)\zeta_1))(d\kappa)^\varpi,$$

$$= \frac{\beta_{\frac{\varpi}{2}}(m_1, m_2)}{(\zeta_2 - \zeta_1)^\varpi}\Phi\left(\frac{\zeta_1 + \zeta_2}{2}\right) - \frac{\Gamma(1 + \varpi)}{(\zeta_2 - \zeta_1)^\varpi}$$

$$\times \left(\frac{1}{\Gamma(1 + \varpi)}\int_0^{\frac{1}{2}}\kappa^{(m_1 - 1)\varpi}(1 - \kappa)^{(m_2 - 1)\varpi}\Phi(\vartheta_1 + \vartheta_2 - (\kappa\zeta_2 + (1 - \kappa)\zeta_1))(d\kappa)^\varpi\right),$$

$$\kappa_2 = \frac{1}{\Gamma(1 + \varpi)}\int_0^{\frac{1}{2}}[\beta_{\kappa\varpi}(m_1, m_2)]\Phi^{(\varpi)}(\vartheta_1 + \vartheta_2 - (\kappa\zeta_1 + (1 - \kappa)\zeta_2))(d\kappa)^\varpi,$$

$$= \frac{\beta_{\frac{\varpi}{2}}(m_1, m_2)}{(\zeta_2 - \zeta_1)^\varpi}\Phi\left(\frac{\zeta_1 + \zeta_2}{2}\right) - \frac{\Gamma(1 + \varpi)}{(\zeta_2 - \zeta_1)^\varpi}$$

$$\times \left(\frac{1}{\Gamma(1 + \varpi)}\int_0^{\frac{1}{2}}\kappa^{(m_1 - 1)\varpi}(1 - \kappa)^{(m_2 - 1)\varpi}\Phi(\vartheta_1 + \vartheta_2 - (\kappa\zeta_1 + (1 - \kappa)\zeta_2))(d\kappa)^\varpi\right),$$

$$\kappa_3 = \frac{1}{\Gamma(1+\varpi)} \int_{\frac{1}{2}}^{1} [\beta_{(1-\kappa)\varpi}(\mathsf{m}_2,\mathsf{m}_1)]\Phi^{(\varpi)}\left(\vartheta_1+\vartheta_2-(\kappa\zeta_2+(1-\kappa)\zeta_1)\right)(d\kappa)^{\varpi},$$

$$= \frac{\beta_{\frac{\varpi}{2}}(\mathsf{m}_2,\mathsf{m}_1)}{(\zeta_2-\zeta_1)^{\varpi}}\Phi\left(\frac{\zeta_1+\zeta_2}{2}\right) - \frac{\Gamma(1+\varpi)}{(\zeta_2-\zeta_1)^{\varpi}}$$

$$\times \left(\frac{1}{\Gamma(1+\varpi)}\int_0^{\frac{1}{2}}\kappa^{(\mathsf{m}_2-1)\varpi}(1-\kappa)^{(\mathsf{m}_1-1)\varpi}\Phi\left(\vartheta_1+\vartheta_2-(\kappa\zeta_2+(1-\kappa)\zeta_1)\right)(d\kappa)^{\varpi}\right),$$

$$\kappa_4 = \frac{1}{\Gamma(1+\varpi)}\int_{\frac{1}{2}}^{1}[-\beta_{(1-\kappa)\varpi}(\mathsf{m}_2,\mathsf{m}_1)]\Phi^{(\varpi)}\left(\vartheta_1+\vartheta_2-(\kappa\zeta_1+(1-\kappa)\zeta_2)\right)(d\kappa)^{\varpi},$$

$$= \frac{\beta_{\frac{\varpi}{2}}(\mathsf{m}_2,\mathsf{m}_1)}{(\zeta_2-\zeta_1)^{\varpi}}\Phi\left(\frac{\zeta_1+\zeta_2}{2}\right) - \frac{\Gamma(1+\varpi)}{(\zeta_2-\zeta_1)^{\varpi}}$$

$$\times \left(\frac{1}{\Gamma(1+\varpi)}\int_0^{\frac{1}{2}}\kappa^{(\mathsf{m}_2-1)\varpi}(1-\kappa)^{(\mathsf{m}_1-1)\varpi}\Phi\left(\vartheta_1+\vartheta_2-(\kappa\zeta_1+(1-\kappa)\zeta_2)\right)(d\kappa)^{\varpi}\right),$$

Thus, by the above expression, the desired identity 66 is obtained. $\qquad\square$

Remark 3.30. *By the change of variable in Lemma, the identity 66 reduces to*

$$\Phi\left(\vartheta_1+\vartheta_2-\frac{\zeta_1+\zeta_2}{2}\right)\beta_{\varpi}(\mathsf{m}_1,\mathsf{m}_2) - \frac{\Gamma(\varpi+1)}{2^{\varpi}}$$

$$\times \frac{1}{(\zeta_2-\zeta_1)^{(\mathsf{m}_1+\mathsf{m}_2-1)\varpi}}\,{}_{\vartheta_1+\vartheta_2-\zeta_2}I^{(\varpi)}_{\vartheta_1+\vartheta_2-\zeta_1}\Omega\Phi$$

$$= \frac{(\zeta_2-\zeta_1)^{\varpi}}{2^{\varpi}}\frac{1}{\Gamma(1+\varpi)}\int_0^{\frac{1}{2}}\left[\beta_{\kappa\varpi}(\mathsf{m}_1,\mathsf{m}_2)+\beta_{\kappa\varpi}(\mathsf{m}_2,\mathsf{m}_1)\right]$$

$$\times \left[\Phi^{(\varpi)}\left(\vartheta_1+\vartheta_2-(\kappa\zeta_2+(1-\kappa)\zeta_1)\right)-\Phi^{(\varpi)}\left(\vartheta_1+\vartheta_2-(\kappa\zeta_1+(1-\kappa)\zeta_2)\right)\right](d\kappa)^{\varpi}.$$
$$\tag{67}$$

Theorem 3.31. *Let $\Phi : [\vartheta_1,\vartheta_2] \to \mathbb{R}$ be a differential mapping on $(\vartheta_1,\vartheta_2)$ with $\zeta_1 < \zeta_2$. If $|\Phi^{(\varpi)}|$ is harmonic convex on $[\vartheta_1,\vartheta_2]$, then the following equalities holds:*

$$\left| \Phi\left(\vartheta_1+\vartheta_2-\frac{\zeta_1+\zeta_2}{2}\right)\beta_{\varpi}(\mathsf{m}_1,\mathsf{m}_2) - \frac{\Gamma(\varpi+1)}{2^{\varpi}} \right.$$

$$\left. \times \frac{1}{(\zeta_2-\zeta_1)^{(\mathsf{m}_1+\mathsf{m}_2-1)\varpi}}\,{}_{\vartheta_1+\vartheta_2-\zeta_2}I^{(\varpi)}_{\vartheta_1+\vartheta_2-\zeta_1}\Omega\Phi \right|$$

$$\leq \frac{(\zeta_2-\zeta_1)^{\varpi}}{2^{\varpi}}\left(\left(\frac{1}{2}\right)^{\varpi}\beta_{\varpi}(\mathsf{m}_1,\mathsf{m}_2) - \beta_{\frac{\varpi}{2}}(\mathsf{m}_1+1,\mathsf{m}_2) - \beta_{\frac{\varpi}{2}}(\mathsf{m}_2+1,\mathsf{m}_1)\right)$$

$$\times \left[|\Phi^{(\varpi)}(\vartheta_1+\vartheta_2-\zeta_1)| + |\Phi^{(\varpi)}(\vartheta_1+\vartheta_2-\zeta_2)|\right]$$
$$\tag{68}$$

for $\mathsf{m}_1,\mathsf{m}_2 > 0$.

Proof. By utilizing Remark 67 with convexity of $|\Phi^{(\varpi)}|$, then we have

$$
\left| \Phi\left(\vartheta_1 + \vartheta_2 - \frac{\zeta_1 + \zeta_2}{2}\right) \beta_\varpi(m_1, m_2) - \frac{\Gamma(\varpi + 1)}{2^\varpi} \right.
$$
$$
\times \frac{1}{(\zeta_2 - \zeta_1)^{(m_1 + m_2 - 1)\varpi}}\, {}_{\vartheta_1 + \vartheta_2 - \zeta_2} I^{(\varpi)}_{\vartheta_1 + \vartheta_2 - \zeta_1} \Omega\Phi \Bigg|
$$

$$
\leq \frac{(\zeta_2 - \zeta_1)^\varpi}{2^\varpi}\left(\frac{1}{\Gamma(1 + \varpi)} \int_0^{\frac{1}{2}} [\beta_{\kappa\varpi}(m_1, m_2) + \beta_{\kappa\varpi}(m_2, m_1)](d\kappa)^\varpi \right)
$$
$$
\times \left| \Phi^{(\varpi)}(\vartheta_1 + \vartheta_2 - \zeta_1)| + |\Phi^{(\varpi)}(\vartheta_1 + \vartheta_2 - \zeta_2)| \right]. \tag{69}
$$

By change variable in the integrals, we get

$$
\frac{1}{\Gamma(1 + \varpi)} \int_0^{\frac{1}{2}} [\beta_{\kappa\varpi}(m_1, m_2) + \beta_{\kappa\varpi}(m_2, m_1)] d(\kappa)^\varpi
$$
$$
= \frac{1}{\Gamma(1 + \varpi)} \int_0^{\frac{1}{2}} \int_0^{\kappa} s^{(m_1 - 1)\varpi}(1 - s)^{(m_2 - 1)\varpi} d(s)^\varpi d(\kappa)^\varpi
$$
$$
+ \frac{1}{\Gamma(1 + \varpi)} \int_0^{\frac{1}{2}} \int_0^{\kappa} s^{(m_2 - 1)\varpi}(1 - s)^{(m_1 - 1)\varpi} d(s)^\varpi d(\kappa)^\varpi
$$
$$
= \left(\frac{1}{2}\right)^\varpi \beta_\varpi(m_1, m_2) - \beta_{\frac{\varpi}{2}}(m_1 + 1, m_2) - \beta_{\frac{\varpi}{2}}(m_2 + 1, m_1). \tag{70}
$$

By writing 70 in the 69, we obtain the required inequality. This completes the proof. $\square$

Remark 3.32. *If we set $\vartheta_1 = \zeta_1$, $\vartheta_2 = \zeta_2$ in Theorem 3.31, then following inequality holds:*

$$
\left| \Phi\left(\frac{\zeta_1 + \zeta_2}{2}\right) \beta_\varpi(m_1, m_2) - \frac{\Gamma(\varpi + 1)}{2^\varpi} \frac{1}{(\zeta_2 - \zeta_1)^{(m_1 + m_2 - 1)\varpi}}\, {}_{\zeta_1} I^{(\varpi)}_{\zeta_2} \Omega\Psi \right|
$$
$$
\leq \frac{(\zeta_2 - \zeta_1)^\varpi}{2^\varpi}\left(\left(\frac{1}{2}\right)^\varpi \beta_\varpi(m_1, m_2) - \beta_{\frac{\varpi}{2}}(m_1 + 1, m_2) - \beta_{\frac{\varpi}{2}}(m_2 + 1, m_1) \right)
$$
$$
\times \left[|\Phi^{(\varpi)}(\zeta_1)| + |\Phi^{(\varpi)}(\zeta_2)| \right] \tag{71}
$$

which is a new inequality in literature.

Remark 3.33. *If in Theorem 3.31, we get $\vartheta_1 = \zeta_1$, $\vartheta_2 = \zeta_2$ and $\varpi = 1$, then, the inequality 3.31 becomes,*

$$
\left| \Phi\left(\frac{\zeta_1 + \zeta_2}{2}\right) \beta(m_1, m_2) - \frac{1}{2(\zeta_2 - \zeta_1)^{(m_1 + m_2 - 1)}} \int_{\zeta_1}^{\zeta_2} \Omega(\varkappa)\Phi(\varkappa) d\varkappa \right|
$$
$$
= \frac{(\zeta_2 - \zeta_1)}{2}\left(\frac{1}{2}\beta(m_1, m_2) - \beta_{\frac{1}{2}}(m_1 + 1, m_2) - \beta_{\frac{1}{2}}(m_2 + 1, m_1) \right) [|\Phi'(\zeta_1)| + |\Phi'(\zeta_2)|]
$$
$$
\tag{72}
$$

which is proved by M. Z. Sarikaya in [26].

Remark 3.34. *If in equation 72, we get* $\mathfrak{m}_1 = 1$, $\mathfrak{m}_2 = \varpi$, *(or* $\mathfrak{m}_1 = \varpi, \mathfrak{m}_2 = 1$*) then the inequality reduces to*

$$\left| \frac{\Phi(\zeta_1) + \Phi(\zeta_2)}{2} - \frac{\Gamma(\varpi + 1)}{2(\zeta_2 - \zeta_1)^\varpi} \left[J^\varpi_{\zeta_1^+} \Phi(\zeta_2) + J^\varpi_{\zeta_2,k} \Phi(\zeta_1) \right] \right|$$
$$= \frac{(\zeta_2 - \zeta_1)}{2^{\varpi+1}(\varpi + 1)} \left[|\Phi'(\zeta_2)| + |\Phi'(\zeta_1)| \right]. \tag{73}$$

which is proved by M. Iqbal in [30].

Remark 3.35. *If in equation 72, we get* $\mathfrak{m}_1 = 1, \mathfrak{m}_2 = \frac{\varpi}{k}$, *(or* $\mathfrak{m}_2 = 1$, $\mathfrak{m}_1 = \frac{\varpi}{k}$*) then the inequalities becomes*

$$\left| \frac{\Phi(\zeta_1) + \Phi(\zeta_2)}{2} - \frac{\Gamma_k(\varpi + k)}{2(\zeta_2 - \zeta_1)^{\frac{\varpi}{k}}} \left[J^\varpi_{\zeta_1^+,k} \Phi(\zeta_2) + J^\varpi_{\zeta_2^-,k} \Phi(\zeta_1) \right] \right|$$
$$= \frac{(\zeta_2 - \zeta_1)}{2^{\frac{\varpi}{k}+1}(\frac{\varpi}{k} + 1)} \left[|\Phi'(\zeta_2)| + |\Phi'(\zeta_1)| \right]. \tag{74}$$

which is given by Sarikaya and Ertugral in [32].

Chapter 4
Conclusion

4 Conclusion

This study investigated relatively new Hermite-Hadamard type inequalities comprising fractal Beta function in Yang's calculus. Based on the theory of neighborhood fractional calculus and generalized convex characteristic within the fractal domain, we provide the main lemma for local fractional integrals. By the use of this most important lemma we set up the inequalities for generalized convex and mercer convex on fractal sets $\mathbb{R}^{\varpi}$ ($0 < \varpi \leq 1$). The results of this analysis is the basis for many inequalities in pure and applied sciences. Furthermore, we demonstrated that the newly discovered inequalities are strong generalizations of similar findings in the literature. Using a new methodology, we utilized Hölder's and Young's integral inequalities to broaden the analysis of Hermite-Hadamard type integral inequalities. It is interesting to extend such findings for other convexities. Authors assume that our recently introduced essence and assumption will be the focus of extensive research in this fascinating field of analysis. Our ambition for the future is to continue in this direction with our research.

Chapter 5
References

References

[1] D. S. Mitrinović, J. E. Pečarić and A. M. Fink, Classical and New Inequalities in Analysis. Mathematics and its Applications (East European Series), *61. Kluwer Academic Publishers Group, Dordrecht*, (1993).

[2] S. S. Dragomir and C. E. M. Pearce, Selected Topics on Hermite-Hadamard Inequalities and Applications, *RGMIA Monographs, Victoria University*, (2000).

[3] P. Agarwal, S. S. Dragomir, M. Jleli and B. Samet, *Advances in Mathematical Inequalities and Applications*, Springer Singapore, (2018).

[4] R. Diaz and E. Pariguan, *On Hypergeometric Functions and Pochhammer k-symbol*, Divulg. Math., 15 (2004), 179-192.

[5] S. Das, *Functional Fractional Calculus*, Springer, Berlin (2011). ISBN. 978-3-642-20545-3.

[6] X. J. Yang, *Advanced Local Fractional Calculus and Its Applications*, World Science Publisher, New York, (2012).

[7] X. J. Yang and J. A. Machado, *A New Fractal Nonlinear Burgers Equation Arising in the Acoustic Signals Propagation*, Math.Methods Appl. Sci., 42(18) (2019), 7539–7544.

[8] X. J. Yang, J. A. Tenreiro and D. Baleanu, *On Exact Traveling-wave Solutions for Local Fractional Korteweg-de Vries Equation*. Chaos, Interdiscip. J. Nonlinear Sci., 26(8) (2016).

[9] X. J. Yang, F. Gao and H. M. Srivastava, *A New Computational Approach for Solving Nonlinear Local Fractional PDEs*, J.Comput. Appl. Math., 339 (2018), 285–296.

[10] A. Carpinteri, B. China and P. Cornetti, *Static-kinematic Duality and the Principle of Virtual Work in the Mechanics of Fractal Media*, Computer Methods in Applied Mechanices and Engoneering, 191 (2001), 3–19.

[11] Y. Zhao, D. F. Cheng and X. J. Yang, *Approximation Solutions for Local Fractional Schrordinger Equation in the One-dimensional Cantorian System*, Advances in Mathematical Physics, (2013), Article ID 291386.

[12] H. Mo and X. Sui. *Generalized-Convex Functions on Fractal Sets*, Abstract and applied analysis, 28(2014), 225–232.

[13] X. J. Yang, *Local Fractional Functional Analysis and Its Applications*, Hong Kong, Asian Academic Publisher Limited. (2011).

[14] G. Jumarie, *Table of some Basic Fractional Calculus Formulae Derived from a Modied Riemann-Liouville Derivative for Non-Differentiable Functions*, Appl. Math. Lett., (22)(2009), 378385.

[15] S. Qaisar, J. Nasir, S. I. Butt and S. Hussain, *On Some Fractional Integral Inequalities of Hermite-Hadamards Type through Convexity*, Symmetry. MDPI., 11 (2019), Article 137.

[16] S. I. Butt, S. Yousaf, A. O. Akdemir and M. A. Dokuyucu, *New Hadamard-type Integral Inequalities Via a General Form of Fractional Integral Operators*, Chaos, Solitons and Fractals, 148 (2021), 111025.

[17] E. Set, S. I. Butt, A. O. Akdemir, A. Karaoglan and T. Abdeljawad, *New Integral Inequalities for Differentiable Fonvex functions via Atangana-Baleanu Fractional Integral Operators*, Chaos, Solitons and Fractals, 143 (2021), 110554.

[18] A. O. Akdemir, E. Set, M. E. Ö zdemir and C. Yildiz, *On Some New Inequalities of Hadamard type for h-Convex Functions*, American Inst. Phys. AIP Conference Proceedings, 1470 (2012) 35–38.

[19] A. McD. Mercer, *A Variant of Jensens Inequality*, J. Ineq. Pure and Appl. Math., 4(4) (73), (2003).

[20] M. Niezgoda, *A Generalization of Mercer's result on Convex Functions*, Nonlinear Analysis, (71) 277, (2009).

[21] L. Horváth, *Some Notes on Jensen-Mercer's type Inequalities, Extensions and Refinements with Applications*. Math. Inequal. Appl., 24(4) (2021), 1093-1111.

[22] J. B. Liu, S. I. Butt, J. Nasir, A. Aslam, A. Fahad and J. Soontharanon, *Jensen-Mercer Variant of Hermite-Hadamard type Inequalities Via Atangana-Baleanu Fractional Operator*, AIMS Mathematics, 7(2) (2021), 2123–2141.

[23] S. I. Butt , S. Yousaf , H. Ahmad and T. A. Nofal, *Jensen-Mercer Inequality And Related Results In The Fractal Sense With Applications*, Fractals, 8(9) (2021), 2240008.

[24] P. Xu, S. I. Butt, S. Yousaf, A. Aslam and T. J. Zia, *Generalized Fractal Jensen-Mercer and Hermite-Mercer type Inequalities via h-Convex Functions Involving MittagLeffler kernel*, Alexandria Engineering Journal, 61(6) (2022), 4837–46.

[25] G. S. Chen, *A Generalized Young Inequality and Some new Results on Fractal Space*, arXiv preprint, (2011).

[26] M. Z. Sarikaya and A. T. A. Fatih, *On the Generalized Hermite-Hadamard Inequalities Involving Beta Function*, Konuralp Journal of Mathematics (KJM), (9)(1), 112–118.

[27] M. Z. Sarikaya, E. Set, H. Yaldiz and N. Basak, *Hermite-Hadamard's Inequalities for Fractional Integrals and Related Fractional Inequalities*, Mathematical and Computer Modelling, (57) (2013), 2403-2407.

[28] R. Hussain, A. Ali, G. Gulshan, A. Latif and M. Muddassar, *Generalized Co-ordinated Integral Inequalities for Convex Functions by way of k -Fractional Deriva-tives*, Miskolc Mathematical Notesa Publications of the university of Miskolc, (Sub-mitted).

[29] S. S. Dragomir and R.P. Agarwal, *Two Inequalities for Differentiable Mappings and Applications to Special means of real numbers and to Trapezoidal Formula*, Appl. Math. lett., 11(5) (1998), 91–95.

[30] M. Iqbal, M. I. Bhatti and K. Nazeer, *Generalization of Inequalities Analogous to Hermite-Hadamard Inequality via Fractional Integrals*, Bull. Korean Math. Soc, 3(52) (2015), 707-716.

[31] M. E. Özdemir, S. S. Dragomir and C. Yildiz, *The Hadamard Inequality for Convex Function via Fractional Integrals*, Acta Math. Sci., 33(5) (2013), 1293–1299.

[32] M. Z. Sarikaya and F. Ertugral, *On the Generalized Hermite-Hadamard Inequalities*, Accepted in Annals of the University of Craiova - Mathematics and Computer Science Series, (2019).

BEI GRIN MACHT SICH IHR WISSEN BEZAHLT

- Wir veröffentlichen Ihre Hausarbeit,
  Bachelor- und Masterarbeit

- Ihr eigenes eBook und Buch -
  weltweit in allen wichtigen Shops

- Verdienen Sie an jedem Verkauf

Jetzt bei www.GRIN.com hochladen
und kostenlos publizieren